Books by Philip J. Klass

UFOs—IDENTIFIED
SECRET SENTRIES IN SPACE

SECRET
SENTRIES
IN
SPACE

SECRET SENTRIES IN SPACE

by Philip J. Klass

RANDOM HOUSE   NEW YORK

To those whose efforts and ingenuity
    have created the remarkable satellites
        that have become a powerful force
            for international stability and peace,
        and whose accomplishments have until now
        been cloaked in official secrecy,
    this book is appreciatively dedicated

My appreciation also
    to my mother, Mrs. Raymond N. Klass,
            for her valuable editorial assistance

CONTENTS

INTRODUCTION

A giant Russian rocket is being prepared for launch as you read this, at a site near Tyuratam, 1,300 miles southeast of Moscow, or from a facility near Plesetsk, 600 miles north of the Soviet capital. If the rocket were loaded with an H-bomb, it could impact anywhere in the U.S. within 30 minutes after launch and would obliterate a major metropolitan area.

But this rocket has quite a different mission—one that is likely to prevent a thermonuclear holocaust. This Russian rocket, and those that will follow it at roughly two-week intervals, will orbit a satellite to make photographs of America's strategic military facilities and those of West Europe and Communist China.

Approximately 90 minutes after launch, as the Soviet satellite completes its first revolution around the earth, giant antennas in western Russia will communicate with the spacecraft to determine if it is functioning properly. During the next revolution, Soviet radars and computers will calculate the satellite's orbital characteristics, determine the precise instant when the spacecraft will be over targets of interest, and instructions will be radioed to the satellite.

During the third orbit (if launched from Tyuratam), as

the satellite passes over Spain at a speed of approximately 18,000 mph, its camera may come to life and take a picture of the U.S. Air Force base near Torrejon. Moments later, the Soviet spacecraft may photograph new underground missile silos which are being constructed in southern France. (If the satellite is launched from Plesetsk, its initial orbits will take the spacecraft over a slightly different ground track.)

Nearly nine hours after launch, late in the sixth orbit, the Russian satellite will begin to pass over the eastern edge of the United States, and it may photograph naval bases at Norfolk and Philadelphia to determine which military ships are in port for resupply. During the seventh orbit, the Soviet satellite will be in position to photograph Minuteman ICBM missile facilities located in Missouri. During the next orbit, the satellite will pass over other missile bases in Montana.

After approximately eight days in orbit, the satellite, if launched from Tyuratam, will have passed directly over all portions of the earth—from the lower tip of South America to the southern flank of Alaska. If launched into a more northerly orbit from Plesetsk, the Russian satellite will have passed over most of North America, including Alaska. Now, as it begins to swing over the lower tip of South America, a computer will initiate the electrical command to fire an onboard rocket, causing the satellite to suddenly slow down and lose altitude. As the vehicle begins to enter the atmosphere, the more dense air will provide further braking action. Finally, as the craft passes over southern Russia at an altitude of about 50,000 feet, a parachute will suddenly pop out and the craft will begin to drift to earth. Once down, the satellite will automatically transmit a coded radio signal to help the Russian recovery team locate it.

A film cassette, containing hundreds of photos, will be removed from the spacecraft and flown by courier aircraft for special processing, probably near Moscow. Then, for weeks, expert photo-analysts will pore over the pictures. They will be looking for any signs of new U.S. missile sites, for details on a new U.S. anti-ballistic missile system being tested on Kwajalein Atoll in the Pacific, and for evidence of Communist Chinese progress in long-range ballistic missiles.

During the eight-day flight, the Russian spacecraft probably will have passed within a few hundred miles of an American satellite, *busily engaged in the same type of "secret" mission*. The U.S. satellite would have been launched from Vandenberg Air Force Base in Southern California.

When Defense Secretary Melvin R. Laird disclosed, as he did in a speech given in New York City on April 20, 1970, that the Soviet Union then had 220 operational sites for its large SS-9 intercontinental ballistic missile (ICBM) and another 60 were under active construction, this was no glib "guestimate." The count has come from a careful analysis of photographs taken by American reconnaissance satellites. On July 10, 1970, Pentagon officials were able to disclose that the Russians had begun to construct additional SS-9 missile sites between June 24 and July 8, providing an indication of how rapidly new strategic intelligence information can be obtained from satellites.

The existence, and purpose, of the clandestine spaceships "that pass in the night" have long been known to officials in both the U.S. and the USSR. The fact that neither nation, especially the Soviet Union, has attempted to destroy the reconnaissance satellites of the other attests to their mutual benefit. This tacit understanding also shows the recognition by both superpowers that the thermonuclear-tipped ICBM, which can destroy the vitals of any nation in a matter of minutes and possibly alter the earth's environment for centuries, has drastically changed the ground rules under which the game of geopolitics must be played.

For decades, Russia had zealously equated secrecy with its own security, and until quite recently there was justification for this view. In July 1955, Soviet leaders understandably rejected President Eisenhower's proposal for an Open Skies aerial-inspection arrangement, believing it was largely an attempt by the U.S. to pinpoint Russian targets for American long-range bombers. Five years later, when the Soviet Union ended U.S. reconnaissance over-flights by shooting down Francis Gary Powers and his U-2 airplane, Premier Khrushchev bitterly denounced the penetrations of the Iron Curtain and used the incident to torpedo the Paris summit conference.

Less than a decade later, American and Russian recon-

naissance satellites* would permit the start of the Strategic
Arms Limitation Talks (SALT) with the Soviets in Helsinki,
and there would be more progress during the next few months
than had been possible in nearly twenty years of previous
discussions. Without reconnaissance satellites, no viable stra-
tegic-arms agreement would be possible because neither
country would allow the other's observers to have unlimited
access to its military facilities. During the SALT negotiations,
Soviet and American delegates will need to discuss what details can or cannot be determined by satellite photos. Perhaps
the negotiators may even exchange "top-secret" satellite photos
to help resolve such issues.

Russian leaders have known of the U.S. reconnaissance-
satellite program at least since 1957, when the first limited
details appeared in the American press. During the next several years, there were guarded references to the project in
censored proceedings of Congressional hearings. By that time,
the project was known as Samos—an acronym derived from
Satellite And Missile Observation System. (Samos is also the
name of a Greek island in the Aegean.)

A Soviet book on military strategy which appeared in early
1962 contained considerable detail on Samos, obviously obtained from U.S. publications. During the summer of 1963,
in Khrushchev's private talks with a West European diplomat,
he referred to the U.S. reconnaissance satellites and jokingly
added that he might let his visitor see some pictures taken by
Russian satellites—but he didn't. In the spring of 1964, when
Khrushchev was interviewed by a former U.S. senator, the
Soviet leader urged the U.S. to end its "provocative" aircraft
reconnaissance flights over Cuba and to substitute American
satellites.

It is ironic that while the Soviets are fully aware of the
American reconnaissance-satellite program, and its effectiveness, the U.S. public has been denied even the barest details

---

* The term "spy satellite" is a favorite with the press, especially
with newspaper headline writers. But it is not an accurate description,
since neither the U.S. nor the USSR has attempted to disguise the
mission of such satellites. The term will be avoided here, generally, in
favor of "reconnaissance satellite."

of the program by its own government. In the fall of 1961, as the Samos satellites began to achieve an operational status, the Kennedy administration dropped a very heavy cloak of secrecy around the program. Even the name "Samos" could no longer be used in government publications. It became a "non-word" in the best tradition of the George Orwell book *1984.*

There was some justification for the original decision in 1961. It had been little more than a year since the Powers U-2 incident, and the administration did not want to offend Soviet sensibilities and provoke Russia into developing weapons to destroy our satellites. Today, however, official secrecy persists because of a general policy that intelligence sources are never discussed publicly. And so, except for an occasional passing reference in the press, most of the American public is unaware of the vital role of reconnaissance satellites in stabilizing relations between the U.S. and the USSR, and in significantly reducing American defense expenditures. Instead, a growing segment of the public has come to believe that the U.S. space program has been no more than a multibillion-dollar "moon-doggle" which is "irrelevant" to the world's major problems.

A clue to the import of the secret part of the American space program was provided by President Lyndon Johnson on March 15, 1967, when he spoke to a small group of educators and government officials in Nashville, Tenn., in remarks intended to be off-the-record. The President said: "I wouldn't want to be quoted on this, but we've spent $35-40 billion on the space program.* And if nothing else had come out of it except the knowledge we've gained from space photography, it would be worth ten times what the whole program has cost. Because, tonight, we know how many missiles the enemy has. And, it turns out, our [previous] guesses were way off. We were doing things we didn't need to do. We were building

---

* The $35-40 billion figure used by the President in 1967 included both the military and civil space programs, such as the Apollo lunar-landing project. Of this total, the military space effort represented about 20 percent. The reconnaissance-satellite program is estimated to have cost approximately $10-12 billion through 1970.

things we didn't need to build. We were harboring fears we didn't need to harbor."*

This book is the story of the U.S. and Soviet reconnaissance-satellite programs, and their impact on world affairs. It is the ironic story of how the mating of two terrifying weapons—the H-bomb and the intercontinental ballistic missile (ICBM)—produced peaceful offspring to hold these fearful weapons in check. It is the story of mankind's most significant benefit from the Space Age. *It is time that the story was told.*

—Philip J. Klass

---

* From *The Nuclear Years,* by Chalmers M. Roberts, Praeger Publishers, 1970.

SECRET
SENTRIES
IN
SPACE

1

MASSIVE RETALIATION–
A SAGGING DETERRENT

"The way to deter aggression is for the free community to be willing and able to respond vigorously at places and with means of its own choosing." In these words, Secretary of State John Foster Dulles first enunciated the new Eisenhower administration foreign policy of Massive Retaliation in a speech before the Council on Foreign Relations in New York City on January 12, 1954. Dulles spoke of the "deterrent of massive retaliatory power," a euphemistic reference to the nation's arsenal of nuclear weapons and its powerful long-range bomber force, the Strategic Air Command (SAC).

The new foreign policy was an outgrowth of the frustrations of the Korean War. Not a few U.S. citizens believed that if nuclear weapons had been employed, or their use threatened, this would have yielded a clear-cut victory instead of a fruitless truce. There was widespread conviction that the nation should never again attempt to wage conventional (nonnuclear) war on the ground against the "Communist hordes." In early 1954, Massive Retaliation seemed to be ideally suited to the military strengths, and weaknesses, of the United States vis-à-vis the Soviet Union and Communist Asia. The U.S. had an overwhelming advantage both in nuclear weapons and in the long-range bombers needed to make the new foreign policy credible. Furthermore, the U.S. seemed likely to keep this edge for the foreseeable future.

Goaded by the prospect that the Korean War might expand into a general war with the Soviet Union, the U.S. had earlier launched a major expansion of facilities for producing the fissionable material needed for atomic weapons. Beyond enlarging those built during World War II at Oak Ridge, Tenn., and Hanford, Wash., new plants were nearing completion at Paducah, Ky., Portsmouth, Ohio, and at Savannah River, S.C. Although these new and expanded facilities would not make their impact felt on the nuclear stockpile for another year or two, the era of "nuclear plenty" was close at hand as Dulles spoke.

True, the Soviets had exploded their first A-bomb on August 29, 1949, several years before expected by most U.S. experts. But in many quarters, this early achievement was credited more to the success of Russian espionage than to native technical skills. In terms of modern technology, the Russians were still considered by many Western observers to be a backward nation. By early 1954 the Russians had tested only five nuclear devices, compared with the 45 exploded by the U.S. This indicated that the Soviets were lacking in the costly, complex facilities needed to produce fissionable material on a large scale.

There was one small cloud on the horizon, however. Five months before Dulles spoke, the Soviets had exploded their first thermonuclear device—the forerunner of an H-bomb whose destructive power could be a thousand times, or more, that of the first A-bombs that had devastated Hiroshima and Nagasaki. The U.S. had tested its first thermonuclear device less than a year earlier, on October 31, 1952. Thus, in the field of nuclear-weapons technology, the Soviets seemed to be closing the gap.

But to deliver such weapons in a massive, rapid-fire attack, the Russians would need a large force of long-range bombers. Here the U.S. appeared to have at least a five-year lead. The USAF's Strategic Air Command was already replacing its World War II vintage bombers, the 300-mph piston-engine B-29s and B-50s with 600-mph pure-jet B-47s. At the time that Dulles spoke, SAC already had a fleet of nearly 600 of the jet B-47s in operation and the number would rise to nearly 2,000 within several years. The B-47s had a range of slightly more than 3,000 miles, but the USAF had developed aerial refueling techniques, using "tanker aircraft," to extend the reach of the B-47. Additionally, by stationing the B-47s at overseas bases in Britain, North Africa, Greenland and in the Pacific, SAC could strike at most of the Soviet Union's industrial vitals and population centers.

A much larger, longer-range jet bomber, the B-52, was already flying in prototype form as Dulles spoke. Production models would begin to enter the SAC inventory by late 1955. In addition to carrying a much larger bomb load, the B-52 had the endurance to strike directly from bases in the U.S.

A single H-bomb carried by one B-47 had the TNT-equivalent destructive power, theoretically at least, of *all* of the conventional bombs dropped by *all* of the participants during the six years of World War II. And a B-52 could carry *several* such weapons! As Dulles spoke, he knew that the U.S. had, or soon would have, the power to completely destroy the Soviet Union and Communist China as viable nations.

The Russian strategic bomber force, so far as was known, consisted entirely of a few hundred 300-mph Tupolev TU-4s, a carbon copy of the U.S.'s old B-29 design. The Soviets had gotten their hands on a few B-29s which had been forced to make emergency landings in Siberia after raids on Japan. The 3,000-mile maximum range of the TU-4 was much too short for a strike against the U.S., except from bases in Siberia which might enable it to reach the Pacific Northwest if unopposed by U.S. air defenders. The U.S. and Canada already were building a $500-million radar network across the Arctic wasteland which would become operational the following year. This Distant Early Warning (DEW) Line could provide at least *four hours'* advance warning of any TU-4 bombing raid. That would be more than adequate to launch U.S. and Canadian interceptor aircraft and, more important, to dispatch most of the SAC bomber fleet against assigned targets in the USSR.

The first Soviet nuclear test in 1949 had jolted the U.S. and Canada into a major, multibillion-dollar effort to bolster their air defenses. By 1954 a vast network of air-defense radars was being deployed within both countries. Giant high-speed digital computers were being built for use in air-defense centers to speed the dispatching and directing of interceptor aircraft to their targets. A new type of automatic all-weather interceptor, the F-102, jammed with electronic equipment, was under development. When it became operational, the air-defense computers on the ground, supplied by radar data on enemy-bomber positions, could instantly calculate the optimum intercept path. The computers then could, by means of a radio-link, take over control of individual F-102s and fly them toward their targets. When the interceptor had closed to within a few miles of its assigned target, its own airborne radar would lock onto the enemy bomber and automatically fly the interceptor into position to launch its guided missiles.

In this context, the new Eisenhower-administration policy of Massive Retaliation seemed like strong armor on January 12, 1954. Yet within several weeks, the first chinks would begin to appear to undermine some of the basic military premises.

On February 15, 1954, the first photographs of two new Soviet prop-jet bombers were published in the U.S. by *Aviation Week* magazine. One was the four-engine Ilyushin IL-38, roughly the size of the U.S.'s new B-52, but somewhat slower in speed. Its range was estimated at 3,000 miles. The other was a six-engine Tupolev TU-200, a 200-foot-long giant comparable in size to the U.S.'s B-36, an early post-World War II bomber. The TU-200 was credited with a range of 4,800 miles, sufficient to enable it to reach many industrial areas within the U.S. on a one-way suicide mission, or with aerial refueling. The magazine did not reveal how it had obtained the photos, but the picture of the TU-200, with landing gear extended, indicated that it had been taken as the bomber was about to land.

A month later, in the March 15 issue, *Aviation Week* reported that the giant TU-200 was believed to be already in squadron service and that estimates of the total number produced ranged from less than a hundred to several hundred.

No source was given for this estimate. Presumably it came from USAF officials who were concerned over the emerging Russian-bomber threat and the fact that the Eisenhower administration was holding down defense spending in an effort to achieve a balanced budget.

On May 1, 1954, the Soviets themselves provided another jolt to the premises on which Massive Retaliation was based. During their May Day parade of military might, a brand-new swept-wing pure-jet bomber, equivalent to the U.S.'s newest B-52, led an aerial fly-by of 175 jet aircraft that also included a squadron of short-range swept-wing bombers. The new Russian long-range bomber, the M-4, with its 600-mph speed, slashed by one half the warning time expected from the new DEW Line radar network—from four hours to a mere two. The U.S. had two B-52 prototypes flying, and the first production model had been rolled out of the Boeing plant in Seattle only two months earlier. The Russians had at least one M-4. How many others they had built U.S. intelligence agencies could only conjecture.

The Eisenhower administration maintained an official silence at first about the new evidence of Soviet strategic air-power. But within the Pentagon and the Central Intelligence Agency there was deep concern. Perhaps the best evidence of this is that before the year was out, the administration had approved plans for Lockheed to build the super-secret U-2 airplane for the very daring assignment of photo reconnaissance over the Soviet Union. However, it would be mid-1956 before the U-2 would begin its first limited surveillance over the fringes of Russia.

In April 1955, a large contingent of U.S. observers converged on Moscow to see what the Russians would unveil during their May Day fly-by. The Russians already were busy rehearsing for the event, and U.S. observers were shocked to see a formation of *ten* of the new M-4 jet bombers in flight. This indicated that the Russians were now in full production on the new jet bomber. At that point in time, Boeing had turned out approximately thirty of its B-52s. The Soviets had at least ten of their new M-4s. Were there more? The Russians also flew a formation of nine of the IL-38 prop-jet bombers, indicating that these too were in production. Observers also saw a formation of thirty new jet fighters which appeared to be equipped with airborne radar to enable them to hunt down SAC bombers during bad-visibility conditions.

For the next two weeks, the Eisenhower administration secretly debated whether to comment publicly on Russian airpower advances. Finally, on May 13, the Defense Department issued the following brief statement:

> The Soviets have recently elected to expose some new aircraft developments in air parade formation over Moscow. These observations establish a *new* basis of our estimate of Soviet production of the heavy jet bomber and of the medium bomber [emphasis added]. There has also been an appearance of a turbo-prop bomber, and a new all-weather fighter has appeared, as expected. This knowledge is evidence of the modern technology of the Soviet aircraft industry and the advances which are being made by them.

This Pentagon statement implied that there was no real cause for concern. But the May 23, 1955, issue of *Aviation Week* reported that the "rapid progress in design and production of supersonic fighters and long-range jet bombers . . . shocked even the top level and most knowledgeable military aviation leaders in the Pentagon."

Democratic Senator Stuart Symington, of Missouri, a key member of the Armed Services Committee, took to the floor to express his grave concern: "It is now clear that the United States, along with the rest of the free world, may have lost control of the air . . . It is now also clear that in quality as well as quantity of planes, the Communists are at least in the process of surpassing the United States—and I am confident they are well ahead with the production of the possible ultimate weapon—the intercontinental ballistic missile . . ." Symington, a former Air Force Secretary in the Truman administration and often a champion of the USAF viewpoint in Congress, called for a Senate investigation. Congressman Dewey Short, a ranking Republican and former chairman of the House Armed Services Committee, called the new disclosures on Russian airpower "somewhat alarming." Congressman George Mahon, a Democrat, spoke of the "seriousness of the situation."

When the issue was raised at the President's next press conference, he sought to play down the administration's concern. But within several weeks, the administration would abandon its efforts to reduce military spending and would authorize a 35 percent increase in the production rate of B-52 bombers. By the end of the following month, it also approved plans to step up production of two new fighter planes, the F-101 and the F-104. Congress was in the mood to respond to the Soviet challenge and it appropriated $6.3 billion in funds for new aircraft for fiscal 1956—more than double the previous year's figure.

During another air show in Moscow on July 3, 1955, the Russians staged a fly-by of twelve of the new M-4 jet bombers. By the end of the year, Air Force intelligence estimates were crediting the Soviets with an M-4 jet-bomber fleet which was nearly twice the size of SAC's own operational inventory of B-52s.

There was another disquieting note—the pace of Soviet nuclear tests. After the first Russian nuclear explosion in 1949, there had been none in 1950 or 1952 and only two each in 1951 and 1953. But during 1954 there had been several—the exact number not disclosed. In 1955 the Soviets staged four nuclear explosions, one of which was the prototype for an H-bomb. The pace would accelerate still more in 1956 with seven nuclear tests, followed by thirteen in 1957.

Even more ominous events were under way in the vicinity of the Ukrainian town of Kapustin Yar, approximately 75 miles east of Volgograd (Stalingrad), near the northwest corner of the Caspian Sea. Giant ballistic missiles were being tested with increased frequency. The consequences would impact heavily on international affairs for many decades—perhaps forever.

2

THE V-2 COMES OF AGE

Almost no one in the West had the vision during the early post-World War II years to foresee that the progeny of the German V-2 ballistic missile would reshape strategic warfare and threaten the international balance of power. There was good reason for this gross misestimate.

The V-2 was a remarkable technological achievement and there appeared to be no defense against the weapon which hurtled down from space at a speed of several thousand miles per hour. But in terms of the warhead weight which the V-2 could deliver, it was an inherently inefficient and costly weapon. The 50-foot high, 30,000-pound ballistic missile could carry less than one ton of explosive. Postwar analyses showed that the 2,400 V-2s launched against Britain and Antwerp did not really cause sufficient casualties and damage to justify the drain on the hard-pressed German economy needed to develop and produce the weapons. The simple German V-1 "buzz bomb," which cost roughly one-hundredth as much, could carry as big a warhead as the V-2. If the Germans had put all their effort into building several hundred thousand of the V-1s, instead of dividing their resources, they could certainly have accomplished far more damage and disruption.

The advent of the atomic bomb seemed to offer promise of enhancing the military value of the ballistic missile, but the weight of early postwar nuclear bombs sorely strained the capacity of the V-2. An even more serious constraint was accuracy. During World War II, the V-2 often missed its aim-point by five miles, or more, after traveling a distance of only 130 miles. Thus, as a possible competitor to the long-range bomber, the ballistic missile seemed to show little promise. Even if it were possible to increase the thrust of its rocket engine by a factor of 7:1 over that of the V-2 to achieve intercontinental range of 5,500 miles, and to improve its guidance accuracy by a factor of 10:1, the missile might still miss its aim-point by 20 miles or more. Even with an atomic warhead, such a weapon was of questionable value, especially when compared with a manned bomber which could drop its bombs within a few hundred yards of its target and, hopefully, return to base for additional missions. The ballistic missile was a single-shot affair.

It is not surprising, therefore, that highly respected U.S. scientists, experienced in modern weaponry, quickly dismissed the prospect of intercontinental ballistic missiles (ICBMs). For example, the brilliant Dr. Vannevar Bush, who directed the nation's research effort during World War II, ridiculed the feasibility of an ICBM during testimony before a Senate committee in December of 1945. "There has been a great deal said about a 3,000 mile high angle rocket," Bush said. "In my opinion, such a thing is impossible . . . (a) rocket shot from one continent to another carrying an atomic bomb, and so directed as to be a precise weapon which would land on a certain target such as this city. I say technically I don't think anybody in the world knows how to do such a thing and I feel confident it will not be done for a very long period of time to come." A group of respected scientists, headed by Dr. Theodore von Karman, convened by the USAF during the early postwar period to recommend promising future technology, reached a similar conclusion.

But in an effort to cover all bets, the USAF awarded a small contract for an ICBM study in April 1946 to Convair *

---

* Now the Convair division of General Dynamics Corp.

(in San Diego). The contract was canceled fifteen months later during an economy move. Instead, the Air Force focused its efforts on long-range bombers and the winged cruise-type missile. The latter was essentially an unmanned bomber, guided by complex electronic equipment.

If the ballistic missile armed with an atomic warhead had a future, it appeared to be as a short-range tactical weapon, to strike at targets up to several hundred miles away. This prompted the Army to sponsor a program at General Electric, called Project Hermes, to recondition and test V-2s which had been captured when Allied troops overran Germany in the closing days of World War II. The first of these V-2s was fired on April 16, 1946, from the Army's White Sands, N.M., Proving Ground, and the last of the 67 missiles was launched on June 30, 1951.

This program was greatly aided by the presence of more than 400 German V-2 engineers and scientists who had headed west just before Germany surrendered to assure that they would end up in Allied rather than Russian hands. The group included Dr. Walter Dornberger, who had directed the German V-2 development and Dr. Wernher von Braun, one of the project's top engineers. Even with the aid of these German scientists, many of whom had worked with liquid-fueled rockets for more than a decade, the V-2 often proved to be a cantankerous beast. This increased doubts that it could ever become a reliable weapon. However, the Army decided to develop a 200-mile improved version of the V-2, large enough to carry an atomic warhead. And so it set up the former German scientists, under Von Braun, at the Redstone Arsenal, in Huntsville, Ala., to develop the first "U.S." ballistic missile, called the Redstone.

Almost nothing was known about Soviet ballistic-missile efforts. The Russians, apparently, had first learned of the V-2 program in the late summer of 1944 after they had overrun Blizna, Poland, where the German Army trained V-2 launch crews. British intelligence, which had begun to piece together the nature of Hitler's new secret weapons, had requested Stalin's approval to visit Blizna. The Russians at first denied the fact that they had yet captured Blizna and the existence of a V-2 missile program. Finally, on September 3, 1944, the Soviets admitted a British intelligence team.

The retreating Germans had carefully stripped the facility, but the British team did recover valuable scrap pieces from nearby craters where malfunctioning V-2s had impacted. These rocket parts were carefully crated in Blizna and the Russians promised to transship them on to London. But when the crates finally arrived in London, they contained only scrap airplane parts, not the original V-2 pieces. Clearly the Russians recognized the import of the scrap which British intelligence was so eager to obtain.

It was known that the Russians had captured some of the German V-2 scientists, but not as many of the top personnel as had come to the West. Presumably the Soviets also had recovered some unused missiles and components from V-2 factories in eastern Germany. In the very early 1950s, the Russians began to release the several hundred German rocket specialists. Western intelligence agents rushed to interview them, to appraise the Soviet ballistic-missile effort. The consensus was that while the Russians were interested in ballistic missiles for short-range tactical use, they appeared to be concentrating on cruise-type missiles for longer-range bombing, apparently for the same reasons that motivated the U.S. to the same policy.

Late in the summer of 1952, a high-level meeting was held at the USAF's Wright-Patterson Air Force Base in Dayton, Ohio, where its Foreign Technology division is located, to consider and appraise information obtained from the returning German rocket scientists. The meeting was attended by representatives from the Central Intelligence Agency, the Atomic Energy Commission and several industrial-scientific organizations. The consensus of those attending was that the Soviets might be able to develop a two-stage missile, using the V-2 base, which could reach the northwest corner of the U.S. and deliver a warhead weighing 2,000 pounds, perhaps by 1956. If so, it would take another two years to stretch the missile's payload to 8,000 pounds and its range to where it could strike any part of the U.S.

One attendee who reportedly felt this appraisal was overly

"optimistic" in terms of likely Soviet progress was the representative from General Dynamics/Convair, whose company had some firsthand experience in the difficult problems of ICBM design. The company once again was working on an ICBM, later to be known as the Atlas. The USAF had awarded a contract in January of 1951 for a modestly funded effort.

Dr. Dornberger had gone back to Germany in 1952 to talk with his former associates who were now returning from Russia. He returned with a more ominous report, that the Soviets were designing a rocket engine with a thrust of more than 260,000 pounds. This was over five times the thrust of the V-2 and more than sufficient for an intermediate-range ballistic missile. But Dornberger's information was considered by some to be self-serving of his own desires to spur a major U.S. program in ICBMs.

What was not known at that moment to any of the returning German scientists, or to those attending the Dayton conference, was that the Soviets had secretly built a *second* ballistic-missile facility, near the one at Khimki, where the captured Germans had worked alongside Soviet engineers. The secret factory was manned exclusively by Russians who carried out a completely independent missile effort. The jointly manned facility was used only as a means of acquiring German know-how and as an independent check on the merit of purely Soviet designs. When the Russians decided that their own rocket technology had surpassed that of the Germans, they sent them home.

The beginning of the U.S. reappraisal of the ICBM's potential as a long-range strategic weapon came in May of 1951. It was the result of laboratory tests at the AEC's Los Alamos, N.M., facility which indicated the basic feasibility of constructing an H-bomb. The H-bomb, with a potential explosive power equivalent to millions of tons of TNT ("megatons") and a thousand times more powerful than the "kilotons" of an A-bomb, could make the ICBM a viable strategic weapon—if other difficult problems could be solved. These included much more accurate guidance systems, more powerful rocket engines, smaller warheads and techniques which could enable the warhead to withstand the turbulent, searing heat of reentry at hypersonic speeds.

By March 1925, some USAF officers were urging acceleration of the Atlas program. Shortly after the AEC demonstrated
a working thermonuclear device on Eniwetok Atoll in the
Pacific, on October 31, 1952, the USAF convened a special
panel of its Advisory Board to consider whether to speed up
the Atlas program. But the panel recommended against such
action because of the many complex problems that still remained to be solved.

The AEC continued its efforts, and by the summer of 1953
its laboratory experiments indicated that the size and weight
of the H-bomb could be reduced drastically, which would
permit a dramatic reduction in the size of an ICBM. The
Pentagon formed a panel of distinguished scientists to consider the implications of this breakthrough for the ICBM. The
group, officially called the Strategic Missiles Evaluation Committee, but better known by its code name, the "Teapot
Committee," was headed by the highly respected Dr. John
von Neumann of the Princeton Institute for Advanced Studies.
The group met for the first time on November 9, 1953, and
three months later it issued its "top-secret" report which urged
an all-out effort to develop and deploy ICBMs. The Eisenhower administration promptly approved.

Seemingly the U.S. had begun to move at top speed as
soon as its best scientists could see any reasonable prospect of
developing ICBMs. Presumably this would insure the continued American preeminence in strategic weapons needed to
deter Communist aggression and to enforce the new Dulles/
Eisenhower foreign policy of Massive Retaliation. Yet a long,
difficult development and test period lay ahead for the U.S.,
and by the following summer (1955) the Soviets would be
testing an intermediate-range ballistic missile (IRBM) capable of hitting targets 1,000 miles away in West Europe. By late
1956 the Soviets would begin to test an even longer-range
ICBM—more than a year before the U.S. would be able to
make its first successful launch of an experimental Atlas.

These disturbing signs of Soviet advantage were known at
the time to U.S. intelligence officials, if not to the U.S. public.
The U.S. had hurriedly developed and installed a giant radar
in the Turkish village of Diyarbakir, near the Black Sea resort
of Samsun. This radar, built by General Electric, began to

track Russian missile launches from Kapustin Yar, starting in mid-1955. The first radar data on Russian missile progress was so disturbing that the National Security Council recommended that the Atlas development effort be given the highest priority in the nation, which was done in September of 1955. A few months earlier, the Eisenhower administration had decided to hedge its bets by authorizing the development of another type of ICBM, called the Titan. This missile, a more advanced design than the Atlas, to be developed by the Martin Co.,* could carry a larger payload over greater distances.

As the radar in Turkey provided growing evidence of Soviet missile progress, the implications were quickly understood by some, but not all, officials of the administration. Recent events in Moscow indicated that the Soviets already seemed to be pulling abreast of the U.S. in long-range jet bombers. If they also should achieve an operational ICBM force a year or two ahead of the U.S., as now seemed likely, the Massive Retaliation foreign policy would be stripped of its deterrent value. Even more ominous, the U.S. itself, for the first time in its history, would lie exposed to a "thermonuclear Pearl Harbor."

The $500-million DEW Line radar network in the Arctic, finished a few months earlier, was completely useless in being able to detect enemy ICBMs coming in from an altitude of several hundred miles at a speed of more than 5,000 mph. The several hours of advanced warning which the Arctic radars were intended to provide, to assure that SAC bombers could become airborne, would soon vanish completely. Thus, a surprise Soviet missile attack could catch practically all of SAC's bombers on the ground, much as the Japanese had destroyed most of the U.S. fleet at Pearl Harbor. But in a thermonuclear-ICBM age, there was one major difference. With SAC's fleet destroyed, the Russians could then use the threat of thermonuclear destruction of U.S. cities to extract any terms they wished. The outlook was grim to those few defense planners who had access to the Turkish radar data.

Even with the highest national priority, the many difficult

---

* Now the Martin Marietta Corp.

technical problems which still needed to be solved ruled out any prospect of accelerating the Atlas timetable significantly. In the fall of 1955, there was a desperate search for counter-measures. One possible stopgap began to emerge in secret conferences within the Pentagon. Sufficient progress had already been made in rocket boosters, missile guidance and warhead design to make it feasible to build an intermediate-range ballistic missile (IRBM), with a range of roughly 1,500 miles. There was hope that the U.S. could develop, build, and quickly deploy such IRBMs at bases in Western Europe where their 1,500-mile range would be sufficient to reach parts of the USSR. Perhaps the IRBMs could hold the fort until the Atlas and Titan could become operational.

On November 15, 1955, acting on the recommendation of the National Security Council, an IRBM project was launched "with a priority equal to the ICBM," but with the proviso that it must not interfere with the ICBM program. Douglas Aircraft was quickly selected to design and build an IRBM, known as Thor, and the Army's Redstone Arsenal, under Von Braun, was authorized to develop another, called Jupiter. The latter would be an enlarged version of the successful Redstone. It was a bold, if costly, effort to counter the approaching Missile Gap.

By early 1957, approximately forty thousand persons in the U.S., in roughly two thousand different companies, were working to close the Missile Gap. The brutal Soviet suppression of the Hungarian uprising only a few months earlier served to underscore the fact that the race was against a ruthless enemy. The dimensions of the tail-chase in which the U.S. was engaged, and the potential consequences of coming in second, were known only to a small number of military officers and civilian officials.

But before the year 1957 was over, unequivocal evidence of the dramatic Soviet lead in long-range ballistic missiles would impact heavily on the American public, and on the world.

3

THE MISSILE GAP IS CONFIRMED

Outwardly the Eisenhower administration maintained an image of calm confidence which belied growing concern in some quarters over the prospective Missile Gap. The President was intent on maintaining "sensible fiscal restraints," and achieving his long-sought balanced budget. The American public seemingly approved because Eisenhower had been re-elected by a larger margin than in 1952. One fleeting indication of concern appeared in the President's State of the Union message of January 10, 1957: "We are willing," the President said, "to enter any reliable agreement which would mutually control the outer space missile [i.e. ICBM and IRBM] and satellite development." The very same day, the Pentagon quietly assigned the highest national priority to all purchase orders required for ballistic-missile production.

On January 25, only fourteen months after work on the Thor IRBM had been authorized, the first experimental missile was poised on its launch pad at Cape Canaveral,* Fla.—a truly remarkable achievement. The missile stood as high as a

---

* Although Cape Canaveral was renamed Cape Kennedy following the President's death, the historic name, Cape Canaveral, will be used here.

seven-story building and weighed approximately 100,000 pounds when fueled. It was powered by one of the two main rocket engines that would be used on the Atlas, capable of producing approximately 150,000 pounds of thrust.

When the long, tedious missile countdown reached zero and the launch button was pushed, the nation's first IRBM burst into flames and exploded ignominiously on the pad! By April 19 a second Thor was fueled and ready for launch. This time the missile got off the pad, but its erratic behavior shortly afterward required that it be destroyed by means of internal explosive charges detonated by radio command. On May 21 a third Thor launch was attempted, but it too proved a dud. For the handful of government officials who had access to the data being obtained from the General Electric radar in Turkey, the three Thor failures underscored the extent of the Soviet lead. The Russians had been launching their IRBMs at the rate of several per month since the previous fall.

U.S. hopes were buoyed on May 31, when the Army Redstone team successfully launched the nation's first IRBM, a Jupiter, which flew a full 1,500 miles. But many, many more Jupiter launches would be needed before all elements of the complex weapon had been debugged and proved. Initial optimism over the first Jupiter success was tempered by radar data from Turkey which indicated that the Russians had begun to test an ICBM prototype, a missile capable of making a direct strike on the U.S.

At the time, the first experimental Atlas ICBM had been cautiously trucked from the Convair factory in San Diego to Cape Canaveral and was being readied for launch. The 80-foot-high Atlas, measuring 10 feet in diameter, weighed 250,000 pounds with its kerosene and liquid-oxygen fuel loaded. At 5 A.M. on June 11, the three-hour-long countdown began. When it reached zero, the giant Atlas began to struggle off the pad and to climb slowly. Then suddenly the plume of flame from its two main rocket engines collapsed and the range safety officer hollered "Losing control." The huge missile began to loop and tumble and headed earthward. The range safety officer pushed the destruct button, and the nation's first ICBM exploded in a ball of fire.

If the unhappy events at Cape Canaveral prompted con-

cern within the Pentagon, they also caused some embarrassment for the House Appropriations Committee. Only a month
earlier a committee report seeking to rationalize the just-
approved fiscal 1958 defense budget had said: "We are no
doubt ahead of the Soviets in the field of guided missiles
generally." The House report conceded that the Russians probably were ahead in the field of short-range IRBMs, but "in the
intercontinental ballistic missile area, we are very probably
ahead of the Soviets."

This illusion was rudely shattered on August 26, when the
Soviet Union announced that it had successfully tested a
"super-long-distance intercontinental multi-stage ballistic missile." The ICBM, the Russians said, had flown at an "unprece-
dented altitude," had covered "a huge distance" and had
"landed in the target area." The Soviet announcement took no
chances that the military import of the achievement might be
overlooked in the West. The tests "show that it is possible to
direct [a] missile into any part of the world." The ICBM,
the Russians added, "will make it possible to reach remote
areas without resorting to a strategic air force, which at the
present time is vulnerable to up-to-date means of anti-aircraft
defense . . ." Finally, to underscore the military significance,
the Soviet announcement added: "A series of explosions of
nuclear and thermonuclear (hydrogen) weapons has been
staged in the U.S.S.R. in recent days. In order to insure the
safety of the population, the explosions were set off at a high
altitude. The tests were successful."

The timing of the Soviet announcement coincided with
East-West disarmament talks which had just gotten under
way in London. The announcement concluded with a reference to disarmament and blamed the West for past failures to
achieve progress in arms-control discussions.

The Soviet ICBM test rated a three-column, three-line
headline on the front page of the August 27 issue of *The New
York Times*. In another front-page story, datelined Washington, the *Times* reported: "The Department of Defense and the
Air Force refused all official comment tonight on Moscow's
announcement that it had successfully tested an intercontinental ballistic missile. It was clear, however, that the announcement would lend new urgency to the United States

program, which already has the highest priority. This was regarded as an inevitable result by military observers here, though they hastened to say that the Soviet statement was not a 'surprise,' a 'shocker,' or 'a cause for alarm' . . . According to United States sources, the Government has had 'ample evidence' for some time that the Russians have been making strides in the development of an intercontinental ballistic missile. But these sources said the fact that the Russians may have successfully test-fired one does not automatically mean that the weapon is ready to go into mass production. Nor does it mean that the balance of power has been upset, or that there has been any decrease in the United States retaliatory power exemplified in the B-52 and B-47 bombers . . . According to these sources, the timing of the announcement, taken in relation to the disarmament talks in London, could be regarded as more psychological than military. They said it might come under the heading of 'blackmail.' " The unidentified military observers, according to the article, "considered it unlikely that the Russians had yet achieved such accuracy despite their uncorroborated statement that they had done so."

More statements intended to reassure were issued in Washington by official government spokesmen on August 27, and reported on the front page of the *Times* the following day under a two-column, two-line headline: "Dulles Says Soviet Missile Will Not Tip Power Scales." The Secretary of State, during his press conference, had said that "he had no reason to doubt Moscow's statement that it had successfully tested an intercontinental ballistic missile. But, he added, he does not think that the military balance of power between East and West will be disturbed by this or other missile developments *for some time to come* [emphasis added]." The Pentagon also issued a formal statement by Acting Secretary of Defense Donald A. Quarles following the Dulles press conference. It said: "Time differences in terms of operational readiness will probably not be very great one way or another, and the immediate significance of the so-called race has been greatly exaggerated."

Administration statements failed to reassure the editorial writers of *The New York Times*. The lead editorial in the August 28 issue noted that "Secretary of State Dulles made no attempt to question the essential validity of the sensational Soviet announcement that Moscow has successfully tested a long-range rocket missile capable of reaching any part of the world . . . it is probably both prudent and correct to assume that the Soviet Union has made a major step forward in rocket and weapons technology.

"We should not try to blind ourselves to the real magnitude of the achievement. We must assume that the Russians have solved successfully the three key problems: creation of powerful rocket motors capable of sending a rocket many thousands of miles, fabrication of a warhead which will not disintegrate from heat before reaching the earth, and development of a navigation system permitting the rocket to be aimed so that it will reach a specific target area . . .

"It is clear that the immediate import of the Soviet achievement is likely to be primarily psychological and political. The Soviet rocket will now make it more possible than ever for the Kremlin to wage a war of propaganda terror against us and our allies, and we may well have much more use of the technique exemplified by last fall's implied threat to bombard France and Britain with rocket weapons [i.e. during the Suez crisis] . . . The free world's statesmen will need stronger nerves than ever. Within our country the Soviet revelation should cause a serious re-examination of past ideas and past policies . . . it is clear that a re-examination of our military policy is required . . ."

It did not take long for Soviet foreign policy pronouncements to reflect the frightening new weapon advance. The August 29 issue of the *Times* carried a story by its Moscow correspondent which began: "With proud references to its newly tested intercontinental ballistic missile, the Soviet Union warned the Western nations today on their stand at the London disarmament talks. It told them they had better meet the Soviet position 'halfway' . . ."

Although Dulles had stated at his August 27 press conference that the U.S. had "no independent means of verification" of the Soviet ICBM claim, within the week the Pentagon officially announced that the Russians had fired not *one* ICBM but "four to six" such missiles during the spring and summer of 1957! Behind the scenes, efforts were spurred to accelerate the U.S. ballistic missile program. On September 20, there was a brief surge of optimism when the first Thor IRBM was successfully launched from Cape Canaveral. But five days later, a second attempt to launch an Atlas ICBM ended in failure after the missile had reached an altitude of roughly 15,000 feet.

If the *Times* clearly perceived the seriousness of the new Soviet threat, much of the nation's press and the public did not. First, there remained a widely held view that Soviet pronouncements should never be taken at face value. Second, the revolutionary impact of the ballistic missile on warfare and geopolitics was unrecognized by most of the public and even by some government officials, both civil and military. The lack of *public* understanding was more excusable since official secrecy cloaked both the extent of Soviet progress and the technical complexities facing the U.S. long-range missile effort.

Secrecy, complacency and lack of understanding vanished in a flash on October 4, 1957, less than six weeks after the Soviet ICBM announcement. On that date, Russia announced that it had successfully launched the earth's first man-made satellite, Sputnik-1, using a powerful ballistic missile. The 184-pound satellite, carrying a scientific payload, was circling the earth every 96.2 minutes at an altitude ranging from 588 miles (apogee) to 142 miles (perigee).

For those knowledgeable in the technology of ballistic missiles, there no longer could be any doubt that the Russians not only had mastered the difficult aerodynamic and propulsion problems of long-range missiles, but that they also had guidance systems adequate to direct ICBMs at the U.S. The significance of the Soviet achievement also was clear to some foreign policy strategists outside the government, such as Dr. Henry A. Kissinger.* In his book *The Necessity For Choice* (New York, Harper & Brothers, 1960), Kissinger wrote: "For the first time, Soviet advances in the missile field—heretofore attributed to the imaginings of overwrought pessimists—became manifest."

If much of the general public failed to fully grasp the

---

* Dr. Kissinger subsequently would become President Richard M. Nixon's special assistant for international affairs.

military implications of the Soviet ability to launch Sputnik-1,
certainly their imaginations were fired by the achievement.
The U.S. press, along with that of the rest of the world, instantly recognized that the Soviet Union had scored both a
major technological and psychological victory. The long-
accepted preeminence of American science and technology
had been rudely challenged.

The lead editorial in the October 7 issue of *The New York
Times* warned that "the Soviet space satellite now revolving
in the skies raises questions of the gravest character regarding
the correctness of our present and past national policies . . .
There has been some tendency in this country these past six
weeks to doubt or minimize the significance of last August's
Soviet claim to have intercontinental ballistic missiles. In the
light of the new evidence, it is the better part of wisdom to
assume that the Soviet Union does have such missile capability, and that it is now engaged in a major effort to provide
itself as soon as possible with large numbers of such weapons.
It is clear we do not have such missiles now, and the time
when we may have them is uncertain."

The *Times* called for the nation's "highest policy makers"
to face up frankly to several vital questions, including: "Is the
world faced with a radical change in the military balance of
power . . . when the Soviet Union has enough such missiles
to place every major United States city and base under threat
of annihilation? Is the policy of putting domestic budgetary
and political considerations ahead of security considerations in
allocating funds for defense still a tenable policy in the present situation?"

Only three days after the launch of Sputnik-1, the Soviet
Union dropped another psychological blockbuster. Moscow
announced that it had tested a new and "mighty" H-bomb,
exploded at great altitude. The giant blast produced shock-
waves that were measured in Japan, but "felt" around the
globe. The next evening, on October 8, during Premier
Khrushchev's visit to the East German embassy in Moscow
he spoke critically of an alleged mobilization of Turkish forces
along the Syrian border. Then he added: "Rockets are terrible
and pitiless weapons and impose heavy responsibility on those
who possess them. We on our part do everything not to permit

war." On the same day, the U.S. asked the United Nations General Assembly to endorse the principle that "outer-space missiles" be brought under international control and that they be used exclusively for scientific and peaceful purposes.

At the President's October 9 press conference, he sought to play down the import of recent Soviet accomplishments. Eisenhower said that they did not increase his apprehensions over the national security of the U.S. "by one iota." But he did express some concern that the U.S. ballistic-missile program was not "further ahead." The President said that if the Soviet satellite really weighed 184 pounds, as the Russians claimed, this achievement "has astonished our scientists."

(More than two years earlier, in the summer of 1955, the U.S. had announced plans to orbit a scientific satellite, called Vanguard, as part of the International Geophysical Year program. The satellite would weigh 22 pounds and measure 20 inches in diameter. Its small size resulted from the government's decision not to use military rocket launchers. Instead, a new civil rocket would be developed so that the Vanguard program could be "unclassified" and open to public view without enabling the Russians to learn about our rocket technology, which in 1955 was thought to be more advanced than that of the USSR. The first attempt to launch a Vanguard would not occur until December 6, 1957, at which time the rocket exploded on the pad. On March 17, 1958, the U.S. finally orbited a miniature Vanguard satellite, the size of a grapefruit and weighing only 3 pounds. Four subsequent launch attempts ended in humiliating failures. Finally, on February 17, 1959, the first 22-pound Vanguard went into orbit, more than fifteen months after the Russians had orbited their 1,120-pound Sputnik-2 on November 3, 1957.)

Despite the efforts of the President to play down the import of the Soviet space achievements, their significance was not lost on America's West European allies. The October 10 issue of the *Times,* in a story datelined Paris, began: "The latest Soviet technical advances call for a critical review of the strategy of the North Atlantic Alliance and for a new kind of leadership by the United States, in the view of European officials intimately associated with the North Atlantic Treaty Organization."

The *Times* article went on to say that "strategy based on ground defenses in Europe and bomber bases in Europe and elsewhere around the territory of the Soviet Union must be reconsidered in the light of the new [Russian] missiles, which may impair the value of such bases . . . The question now asked is: If Moscow threatened to drop long-range hydrogen missiles over a given country, as it threatened Britain and France when they invaded Egypt last December, would the people of that country remain calm in the confidence that the air power of the Atlantic Alliance would protect them?"

Back in the U.S., the chairman of the Senate Armed Services Committee, Richard Russell (Dem., Ga.) said: "We now know beyond a doubt that the Russians have the ultimate weapon—a long-range missile capable of delivering atomic and hydrogen explosives across continents and oceans." Russell added that "this is no time for panic or fright." Senator Stuart Symington said: "If this now-known superiority over the United States develops into supremacy, the position of the Free World will be critical," and he called for a full Senate investigation. Senator Lyndon B. Johnson (Dem., Tex.), chairman of the Preparedness Investigating Subcommittee of the Senate Armed Services Committee, announced plans for a prompt investigation to find out why the Russians had beaten the U.S. in launching a satellite.

A few days later, the Congress and the U.S. public were jolted again. The October 21 issue of *Aviation Week* magazine carried a major story which revealed the existence of the radar in Turkey that had provided hard evidence of Soviet advances in ballistic missiles for two years prior to the official Russian announcement. According to the *Aviation Week* story, the radar data showed "a significant shift from the irregular pattern of experimental test firings to a regular five-per-month pattern . . . during 1956. This provided fairly conclusive evidence that the Soviet IRBM program had shifted from the development phase to production with an operational capability imminent."

The article said that "Detection of longer-range multi-stage ICBM test programs along the 70-degree track toward the Pacific began in late 1956. A variety of shots was recorded, including stage separation tests, maximum altitude attempts and finally long-range firings impacting about 4,000 miles from launching sites . . . These long-range firings began during the early summer of 1957 . . . During the summer months of June, July and August, there were at least eight firings of long-range multi-stage missiles . . . along the Siberian track . . . The long-range missile firings, frequencies, irregular intervals and variety of tests conducted indicate that the Soviet ICBM program is still in the development test stage . . ."

The same issue of the magazine had a hard-hitting editorial, by Editor-in-chief Robert B. Hotz, which sharply criticized the Eisenhower administration for imposing a restrictive ceiling on defense expenditures to achieve a balanced budget. Hotz added that the presence of the giant radar in Turkey was well known to the Russians and that *Aviation Week* had known of its existence for over a year. The Turkish-radar disclosure made headlines around the world and sparked similar editorial criticism in a number of major newspapers around the country. Some referred to the emerging Missile Gap as a "technological Pearl Harbor."

Senator Johnson promptly announced that the Preparedness Investigating Subcommittee would now hold a "searching inquiry into the entire field." The Pentagon quickly arranged a briefing for Congressional leaders which lasted for *seven hours!* When Johnson emerged from the briefing he commented to the press: "Russia is ahead of us . . . The U.S. is lagging in both the satellite and missile programs." Johnson said the Soviet Union "has handed us both a technological and propaganda defeat," and that "feeble denials of the facts of life" would not cure the situation. Johnson cautioned against "panic over our own lagging program."

Even the senior Republican member of the subcommittee emerged in a somber mood from the Pentagon briefing. New Hampshire's Senator Styles Bridges said he was "deeply concerned with the overall progress" of the U.S. missile-satellite program. "We have no time to lose," Bridges warned.

The successful launch of the USAF's second Thor IRBM on October 24 lifted the gloom in Washington only briefly, for less than two weeks later Russia orbited its 1,120-pound

Sputnik-2, nearly four hundred times the weight of the tiny Vanguard which would not go into orbit until the following March.

On November 7 the Soviets celebrated the fortieth anniversary of the Bolshevik revolution with a massive military parade in Moscow. Included was a 70-foot long intermediate-range ballistic missile with a reach estimated at 1,000 miles. As the Soviets were displaying their new ballistic weapons, top Air Force and civilian scientists were meeting secretly in Baltimore at the request of the President. Their purpose was to make a fresh appraisal of the relative military airpower of the U.S. and Russia. Their verdict was more pessimistic than an earlier one that the Soviets could not be in a position to launch a massive ICBM attack before 1960. The consensus of the Baltimore group was that the period of grave Soviet threat might begin as early as the fall of 1958, only a year away.

The closing weeks of 1957 brought one welcome development. On December 17 the first successful launch of an Atlas ICBM was achieved. The giant missile flew only a 500-mile trajectory, intended to test its two main rocket engines and airframe integrity. But at least the "big one" had finally flown. Two days later, the USAF successfully launched its fourth Thor IRBM. It was the first to be equipped with a self-contained "inertial-guidance" system of the type planned for use in operational missiles.

But the closing weeks of 1957 also brought the President a disturbing report from a special committee he had convened to make an impartial outside-of-government appraisal of the emerging missile-space gap and of its consequences. The committee was headed by H. Rowan Gaither, chairman of the board of the Ford Foundation. The committee report had been submitted to the National Security Council late in November, prior to submission to the President.

Despite the "secret" classification of the report, word of some of its conclusions began to leak to the press. According to the resulting news stories, the Gaither Report warned that by 1960 the Strategic Air Command bases around the globe could be wiped out in a matter of minutes by a surprise Soviet missile attack. The report was said to have urged that SAC's bombers be dispersed to a large number of bases to reduce their vulnerability. Further, the Gaither committee reportedly urged a significant increase in defense spending.

The Eisenhower administration refused to make public or discuss the contents of the Gaither committee report on the grounds that its data on Soviet capabilities was "highly classified." This seemed to confirm the ominous-sounding news stories. Britain's *New Statesman* summed up the prevailing view in an editorial in its January 4, 1958, issue in these words: "The Gaither report has revealed that, irrespective of any efforts which America may now make, *the Soviet preponderance in advanced weapons has reached such an absolute state that America's national survival will depend, until 1961 at least, on 'Russian benevolence'* [emphasis added]."*

Less than four years earlier, Secretary of State Dulles had confidently announced the then new U.S. policy of Massive Retaliation to deter Communist expansion. Now the military foundations of that policy appeared to be crumbling rapidly. At some time during the next several years, it seemed almost certain that SAC's deterrent power would diminish drastically. When that happened, the U.S. and the Western world would lie vulnerable to a possible ballistic-missile/thermonuclear Pearl Harbor.

---

* Today, some authors with extremely short memories charge that the Missile Gap was only a myth perpetrated by the "military-industrial complex" in collaboration with "the kept press of the missile and aviation industries," as one author recently phrased it.

4

THE RACE TO CLOSE THE MISSILE GAP

The immediate threat in early 1958 was not quite as grim as it appeared to the *New Statesman*'s editorial writer. SAC's fleet of short-range B-47 bombers had grown to more than 1,500 in number and there now were a few hundred of the longer-range B-52s in the operational inventory. But it had become clear that the ballistic missile was the dominant weapon of the future, and here the Soviets were in the lead.

Top-secret reconnaissance flights of the U-2 over the Soviet Union, which began cautiously in June 1956, had been gradually expanded. By the spring of 1957, a new missile test site for longer-range missiles had been discovered at Tyuratam, near the Aral Sea—roughly 700 miles east of the one at Kapustin Yar. By mid-1957, a U-2 flying out of Pakistan had brought back photos of the new ICBM test site.

The U-2 photos failed to show any evidence of widespread Russian deployment of the new jet bombers which had been displayed so ostentatiously several years earlier. But the pictures did reveal that the Soviets were rapidly expanding their air-defense radar network and numbers of interceptor aircraft and air-defense missiles. The U-2 photos prompted the Central Intelligence Agency to downgrade its earlier estimates of

Soviet bomber strength, made in the spring of 1957. But in CIA testimony before Congress, the agency did not reveal the basis for the new estimates. This prompted some Congressional critics, such as Senator Symington, to conclude that the CIA's action was merely an attempt to support the administration's efforts to hold down defense spending.

The absence of photographic evidence of widespread Soviet deployment of new jet bombers was disquieting to some. If the Russians had decided to by-pass production of their promising new jet bombers, this almost certainly meant that they were plunging ahead with all-out production of IRBMs and ICBMs. Later in the year there would be evidence that the Russians had indeed stepped up their defense spending sharply.

Within the CIA, one analysis of the situation concluded that *if* the Russians were indeed making an all-out production effort to exploit their present lead, they could have as many as 500 of the ICBMs in readiness by the summer of 1960, only two years hence. By that date, the U.S. would have fewer than a dozen ICBMs in place. If the Russians had several hundred ICBMs operational by mid-1960, this would be sufficient to enable them to launch a massive surprise attack which could knock out the handful of U.S. missiles, destroy all of SAC's bases and most of its aircraft, as well as all major U.S. cities. On this basis, the period of U.S. peril would begin to develop in 1959 and probably reach its peak during 1960–61. After 1961, production of U.S. missiles would, hopefully, reduce the frightening disparity.

Even as U.S. scientists and military weaponeers strained to develop, test and debug the Thor, Jupiter, Atlas and Titan, the realization began to emerge that these weapons were already obsolescent even before they had become operational. But this view was closely held because open discussion of the fact would hardly reassure the U.S. public or American allies. The reason was this: a giant, thin-skinned ballistic missile, poised on its launch pad, is a sitting duck for a surprise enemy attack and could be knocked out by the blast from a thermonuclear weapon exploding several miles away. Unlike a bomber fleet, whose aircraft can be shuffled between a variety of air bases or kept on airborne alert, the long-range missile was anchored to its launch pad, with its complex fueling and check-out equipment.

A ponderous liquid-fueled ICBM or IRBM, which required at least 15 to 30 minutes to fuel up and test prior to launch, and which could not remain fueled for extended periods, was a viable weapon only for a nation whose strategy might call for it to launch the first blow, or which could hide such weapons in the vast expanses of its countryside through rigorous internal security measures, as the USSR could do. But in the open U.S. society, the approximate location of all ICBM launch pads would quickly become known to Soviet military planners and could therefore be easily destroyed in the first surprise wave of Russian missiles.

Even before the first experimental Atlas had succeeded in traveling its intended range, U.S. defense planners were taking a second look at the ICBM as a viable weapon against a large and very secretive potential enemy. One means of providing some degree of protection would be to house the ICBM in an underground silo where it might survive all but a direct hit or near miss. If the missile survived, an elevator could bring it to the top of the silo for launch. Plans were developed to build such silos for later versions of the Atlas, and more rugged silos for the follow-on Titan ICBM.

An even more imaginative approach to protecting the thin-skinned ballistic missile from surprise attack also was emerging. When the U.S. had decided to initiate the stopgap program in late 1955, the Navy had briefly considered the possibility of launching IRBMs from surface ships, which could present an ever-moving target to enemy attack. But the problems of launching a 65-foot-high missile from the deck of a pitching, rolling ship, and the risks involved in handling the volatile liquid-oxygen fuel, prompted the Navy to drop the idea in favor of an even more daring concept. This was the idea of launching an IRBM from a submerged nuclear-propelled submarine, a concept that had become possible as the result of recent developments in solid-propellant fueled rockets. On December 17, 1956, the Navy had contracted with Lockheed Aircraft Corp. to explore the idea. Lockheed had gained considerable experience with solid-propellant rockets as a result of its X-17 multistage rocket, built to test the survivability of reentry-vehicle (warhead) designs to be used on operational IRBMs and ICBMs.

If the high-risk idea paid off, the U.S. would gain a missile whose launch pad could continuously roam the depths of the ocean, making it nearly invulnerable to surprise attack. The new submarine-launched missile became known as the Polaris. The Polaris program also sparked the thinking of Air Force planners who were seeking improved ways to protect land-based ICBMs and to reduce the long reaction time of liquid-fueled missiles. If the Polaris could be launched from a submerged submarine, then a similar solid-fueled missile with longer ICBM range could be designed to be launched directly from a deep underground silo.

On February 20, 1958, the Pentagon approved the start of the new Air Force solid-propellant ICBM program, under the code name of Weapon System 133A. Subsequently the missile became known as the Minuteman. Thanks to recent advances in thermonuclear weapon design, the Minuteman could be reduced in size to roughly two-thirds the length of the Atlas and Titan, and would measure 53 feet long and 6 feet in diameter. The smaller Minuteman could be housed in a deep, hardened underground silo and suspended on giant springs to protect the missile from the shock of a nearby explosion. With the new solid-propellant fuel and a new-type guidance system, the Minuteman could be kept on continuous alert, ready to be launched in a matter of a minute or two.

In theory at least, this would enable a U.S. President to order the launch of Minuteman missiles as soon as the first enemy warheads impacted, or even before, if reliable warning of the attack could be obtained in advance of impact. Early in 1958, contracts had been awarded to build two giant radars to provide 15 minutes' advance warning of an enemy ICBM attack. The radars, similar to the one in Turkey, would be built at Thule, Greenland, and in central Alaska. Even if a President declined to exercise the option offered by these early-warning radars, most of the Minuteman missiles were expected to survive a surprise enemy missile attack. The Ballistic Missile Early Warning System (BMEWS) radars also could provide a vital alerting function for SAC bombers, enabling a larger percentage of the fleet to become airborne and avoid being destroyed by a swift ICBM attack. It would cost more than $800 million to build the two giant BMEWS radars in the harsh Arctic, but even at this price they seemed worth every penny invested.

But the first BMEWS radar at Thule was not scheduled to go into operation until the fall of 1960, and it would be the summer of 1961 before the one in Alaska was ready. *Until then, there would not even be 15 minutes' warning of an enemy ICBM attack if it should come.*

If the crash-program timetables that had been set for the Polaris and Minuteman missile developments could be met, the first submarine-launched missiles might become operational in late 1960 or early 1961, with the first Minuteman missiles becoming operational a year later. But if these new weapons encountered even a fraction of the problems experienced by the first-generation liquid-fueled missiles, the Missile Gap was likely to extend well into the mid-1960s—or so it appeared in the spring of 1958. On February 28 an improved Thor IRBM, equipped with the prototype of an operational reentry vehicle and new "vernier engines" designed to improve guidance accuracy, was successfully flown from Cape Canaveral. And on March 26 the Navy successfully fired a "dummy" Polaris missile from a submerged launch platform.

A few days previously, the Soviet Foreign Ministry had finally responded to President Eisenhower's earlier proposal for a ban on the use of outer space for military purposes. The Soviets insisted that any such ban had to be accompanied by "liquidation of foreign military bases in Europe, the Middle East and North Africa." This would have stripped away the U.S. bases needed for shorter-range B-47s and tanker aircraft used for airborne refueling, almost completely emasculating the Strategic Air Command. Clearly the Soviet conditions were unacceptable at a time when the SAC bomber fleet seemed to be the only possible offset to the emerging Soviet superiority in ICBMs.

The first Atlas ICBM equipped with the three-rocket engines which would be used in operational versions (two main engines plus a smaller "sustainer" engine) was successfully launched on July 19 from Cape Canaveral. But it had to be destroyed two minutes later because of a control-system malfunction. On August 2 another attempt was made and this flight was successful. Then, on August 29, 1958, a full-powered Atlas was successfully flown over a 3,000-mile trajectory—approaching intercontinental range. The achievement came approximately one year after the Russians had done so. But another Atlas launch attempt a few days later saw the missile explode 85 seconds after take-off. Clearly the giant brute had not yet been tamed. On November 28, the first fully powered Atlas made a full-range (6,325-mile) flight and landed "close to its target," according to official statements released after the mission. Still, the Atlas had a long way to go to achieve the status of an operational weapon.

Optimism over this first full-range Atlas flight was tempered by an obscure sentence in an 88-page report published during this period by the Soviet Union outlining Russia's new seven-year plan. The sentence read: "Production of intercontinental ballistic rockets has been successfully set afoot." In the months since the August 26, 1957, Soviet announcement of its successful ICBM tests, Premier Khrushchev had on several occasions hinted that ICBM production was under way. But his remarks had been given little credence in the West because of Khrushchev's characteristically boastful ways.

The first attempted launch of the new Titan ICBM on December 20 ended in failure. But three days later the newest Atlas-C version was successfully fired from Cape Canaveral. Meanwhile, early optimism over initial progress on the Navy Polaris program had received a series of setbacks. The first attempt to launch a full-scale missile from a land-based pad at Cape Canaveral on September 24 had seen the missile reach an altitude of 40,000 feet; then it had shuddered and exploded. Another attempted launch in mid-October had resulted in premature ignition of the second stage which soared skyward, leaving the first stage aflame on the launch pad. On December 30 a third Polaris was ready for launch at the Cape. When the countdown reached zero, the missile took off smartly, climbed smoothly for 45 seconds, but then it pitched over erratically and had to be destroyed by the range safety officer.

The Polaris seemed destined to follow the same long, troubled path of its liquid-fueled predecessors. Any hopes that it could be deployed operationally aboard submarines within two years seemed at this point like nothing more than wishful thinking. Several months earlier, the Air Force had selected Boeing Airplane Co. as the prime contractor for the Minuteman silo-based missile.

As 1958 drew to a close, little that had occurred during the year provided much cause for optimism that the U.S. was closing the Missile Gap. Perhaps the Soviet lead was even growing. In the fall of 1958, Secretary of the Air Force James Douglas had publicly admitted that the Russians were expected to have a "small number"of operational ICBMs during 1959, and a "substantial number" in operational squadrons by 1960.

In December, Senator Hubert H. Humphrey (Dem., Minn.) returned from Moscow, where he had had an eight-hour interview with Khrushchev. Humphrey brought back a disquieting report: the Soviet leader had told him that Russia had now developed a super long-range ICBM, capable of carrying a five-megaton warhead over a distance of 8,700 miles, a range sufficient to reach any city within the U.S., as well as all of America's overseas bases. During 1958 the Russians had been especially active in nuclear testing of weapons suitable for use in ICBMs. They had carried out twenty-five nuclear tests, nearly as many in one year as they had conducted during all of the previous decade!

But perhaps the most disquieting event of 1958 had come on November 10, when Premier Khrushchev had announced that Four-Power occupation of Germany must be ended and that West Berlin must be converted to a "free city." Soviet foreign policy was beginning to reflect the USSR's superiority in the "ultimate weapon."

5

THE DECADE OF PERIL APPROACHES

The events of early 1959 did little to alter the pessimism that had pervaded the U.S. since Sputnik-1 went into orbit. On January 19 the Navy made its fourth attempt to launch an experimental Polaris missile which promptly malfunctioned and destroyed itself. A fifth Polaris launch attempt on February 27 also ended in failure.

The Senate Preparedness Investigating Subcommittee launched hearings on January 29 to reappraise the nation's missile/space program. Senator Symington challenged the administration's latest intelligence estimates of Soviet missile strength. He charged that the figures had been "changed radically" from those he had been given the previous summer which were the "best figures that the Central Intelligence Agency had." Even those estimates of Soviet ICBMs, he had felt, were too low, and now the figures had been further reduced despite the elapsed time during which the Russians would be expected to have built more missiles. Although Symington had been Secretary of the Air Force under Truman and therefore had excellent information sources in that service, and although he held a key position on the Senate Armed

Services Committee, he was not one of the select few who had been told of the top-secret U-2 missions over Russia.

What Symington therefore did not know was that U-2 photos obtained during the last half of 1958 had failed to show any indication that the Russians were mass-producing and deploying ICBMs. It was this that had prompted the CIA to revise downward the intelligence estimates of the previous summer. But there were some within the CIA and the USAF intelligence community who disagreed. They argued that the Russians probably were deploying their ICBMs in the more remote parts of the vast Soviet Union over which the U-2 had not yet ventured.

In response to Symington's challenge on the latest intelligence figures, Defense Secretary Neil McElroy could only hint at their rationale, observing that "one of the more difficult things arises from our having to depend for our actions on information which is not available to the public." McElroy admitted that "it is not our intention or our policy to try to match missile for missile in the ICBM category . . . in the next couple years . . . our diversified capability [i.e. bombers plus missiles] to deliver the big weapon is what we are going to count on . . ."

Symington shot back: "What you are in effect saying is that we are planning in this fashion in spite of our intelligence estimates that the Russians are going to produce a great many more missiles, ICBMs, than we are; is that correct?" McElroy replied: "We are planning based on the assumption that they have that capability, and if they use that capability—and we are assuming that they will—then they will have a year from now, two years from now, a larger number [of ICBMs]." But he added that "there is nothing in our intelligence that indicates that operationally he [Russia] is ahead of us at this point." During the hearings, Strategic Air Command chief General Thomas S. Power warned that a surprise Soviet ICBM attack could catch most of SAC's bombers on the ground. Power urged that funds be provided to maintain an "airborne alert" in which a significant percentage of SAC's bombers would be kept airborne, loaded with bombs.

These and other expressions of concern were promptly underscored by events in Moscow. Khrushchev, addressing the 21st Communist Party Congress, spoke of "serial [mass] production of intercontinental ballistic rockets." Soviet Defense Minister Marshal Rodion Malinovsky told the Soviet group that Russia's army "is equipped with a whole series of intercontinental, continental and other rockets of long, medium and short range." He added that Soviet ICBMs carried H-bomb warheads which could hit "precisely any point."

When President Eisenhower was asked about these Soviet boasts at his press conference, he cautioned against taking them at face value. But he admitted that "apparently, they are believed all around the world, and too implicitly." The New York *Herald Tribune* and *Life* magazine, both of which normally supported the Republican administration, apparently were among those who accepted the Soviet claims because both publications carried editorials urging a prompt reexamination of current defense policies to meet the Soviet missile threat. In mid-February, Representative George Mahon (Dem., Tex.), chairman of the House Appropriations Committee, asked McElroy to submit an estimate of the cost of closing the Missile Gap. Democratic Congressional leaders privately discussed the possibility of adding $500–700 million to the administration's $40.8 billion fiscal 1960 defense-budget request to accelerate Atlas and Titan production.

Some U.S. intelligence people gave credence to the report that the Russians had completed tests of their large T-3 ICBM and were now already producing them at the rate of 15 missiles per month. If true, the Russians could have more than 100 ICBMs in operational readiness by the end of 1959, and the total might climb to 600 by the end of 1962. (The T-3 is currently called the SS-6.)

By early March of 1959 the U.S. finally had made two successful launches of the first stage of the two-stage Titan rocket. Even before the first experimental Titan had been flown in its two-stage configuration, construction was already under way on the first of three underground silos in which Titans would be housed at Vandenberg Air Force Base in California. This philosophy of proceeding with launch-site construction even before the missile had been fully developed was a costly approach, but it seemed to be the only possible way of quickly closing the Missile Gap. The administration also was gambling, on the basis of its U-2 reconnaissance, that the Polaris and Minuteman missiles, the promising new second-generation solid-propellant weapons, could be brought along rapidly enough to close the gap by the early 1960s.

During the Senate hearings, CIA director Allen Dulles admitted that the current Russian superiority in ballistic missiles was likely to make the USSR "more assertive." He expressed the opinion that the Soviets hoped "to be able to hold the U.S. under threat of nuclear attack by intercontinental ballistic missiles while they consolidate their position in the fragile parts of the non-communist world."

SAC Commander Power warned, in testimony before the House Defense Appropriations Subcommittee, that U.S. ability to deter Soviet expansion was rapidly deteriorating as a result of the "fantastic compression of warning time and reaction time" in an ICBM era. He described how SAC kept some of its bombers fueled and loaded with bombs near the end of the runways, with their crews living and sleeping in nearby trailers, so that at least a few could become airborne within five minutes after warning was received. But the BMEWS radars in the Arctic, intended to provide the critically needed 15-minute advance warning, would not become operational for more than another year. During Power's testimony, Representative Mahon observed that "nobody can tell unequivocally and with complete assurance just what the situation is within the Soviet Union with respect to their present capability and prospective capability." Power responded: "One of our weakest areas is the little information we have to work on. If there is anything that needs emphasis . . . , it is the ability to get more information about the Russians and particularly their missile capability . . ."

Hopes that the Atlas was nearing operational status were sharply jolted in the spring of 1959, when a series of five missiles in succession exploded in flight. The last three failures, on April 14, May 19 and June 6, were especially vexing because they were the latest design, the Atlas-D, which it had been hoped would be the first operational configuration. Defense Secretary McElroy, in announcing the failures, said the problems had not yet been isolated and that the initial operational deployment of the Atlas would therefore be de-

layed by at least sixty days. Seeking to soften the impact of
the bad news, McElroy said: "We believe that they [Russians]
have had serious trouble. Whether they have worked out of
it we don't know, but I would be inclined to believe from the
intelligence estimates that they will probably just about make
what has been expected of them." This, McElroy said, was
that the Soviets would have about 10 operational ICBMs by
the end of the current year (1959). But he added that the
precise number of Russian missiles was not really important.
"The important thing is when they get enough to 'cream'
[destroy] the country."

The summer of 1959 was not entirely black. On June 9
the first nuclear-powered submarine, the USS *George Washington,* designed to carry Polaris missiles, was launched. During the same month, the U.S. dispatched the first of its Thor
IRBMs to Britain—the stopgap was on its way. From Britain,
the Thor could hit at a number of major cities in western
Russia. Also, the first successful launch of a Titan in which its
second stage was fired and separated from the first-stage
booster was achieved. But an attempt to launch an experimental Polaris ended in failure when its first-stage booster
went out of control.

Khrushchev's earlier demand for making Berlin a "free
city" was being pressed by the Soviets in the "more assertive"
manner that CIA's Dulles had forecast. This suggested that
perhaps Soviet ICBM strength was growing faster than U.S.
intelligence had estimated, notwithstanding the U-2 photos.

The last two weeks of July brought good news that the
troubles that had plagued the Atlas had been traced to a
fuel valve and had now been fixed. Two successful launches
in a row were conducted within a matter of days and one of
these covered the full design range of 6,300 miles. But the
mood of the Congress was reflected in the fiscal 1960 budget
which was completed during the first week in August. Congress decided to authorize a total of 17 Atlas missile squadrons, nearly twice the number sought by the administration.
And an extra $77 million was voted to accelerate the Minuteman missile program.

In late August the Russian Navy's official newspaper published a warning that Soviet missile-launching submarines were now able to enter Hudson Bay under the Arctic ice and could fire their weapons at U.S. industrial targets 700 miles away. Only the day before, the U.S. Chief of Naval Operations, Admiral Arleigh A. Burke, had predicted that the Russians might soon have this capability.

On September 1, after four successful Atlas launches in a row, the USAF announced that the nation's first ICBM was now "operational" and a handful of launch pads at Vandenberg AFB, Calif., were officially turned over to SAC. (Fortunately the "operational status" of those early Atlas missiles was never put to the test.) Eight days later, a SAC crew fired its first Atlas missile from Vandenberg and its unarmed warhead (reentry vehicle) impacted 4,480 miles away, near Wake Island. A few days earlier, the Navy had successfully launched a Polaris from a ship, the USS *Compass Island,* configured to simulate a missile launch from a submarine. It had come less than five months after the Navy finally had succeeded in launching its first experimental Polaris from a land-based site, on April 6. After a painful start, the Polaris program was beginning to move.

On September 12, Premier Khrushchev arrived in the U.S. for a good-will meeting with President Eisenhower and tour of the country. On the same day the Russians launched an 858-pound spacecraft which became the first man-made object to impact on the moon. The timing of the lunar shot, Khrushchev said, was "a simple but pleasant coincidence." But to the editorial writer of the Danish newspaper *Information,* the Soviet feat had more significant implications: ". . . now we know that an H-bomb-carrying rocket can with precision hit New York," he wrote.

A few days later, the Navy successfully launched a Polaris from Cape Canaveral which covered 900 miles, nearly the full design range. Within the next several weeks, an Atlas-D equipped with an operational-type reentry vehicle (minus its thermonuclear bomb) successfully made a 5,000-mile flight. During the first week of November the U.S. formally announced plans to deploy a squadron of Jupiter IRBMs in Turkey, to be operated "initially" by USAF personnel. Two other Jupiter IRBM squadrons were being readied for deployment in Italy as well.

As 1959 approached its closing days, the U.S. strategic missile program seemed, at long last, to be making mildly encouraging progress. The Titan, with its more advanced two-stage design, was experiencing some difficulties, as had the Atlas a year earlier. Since the last successful Titan flight on May 4 there had been only failures, including the final attempt of the year in mid-December, when the missile exploded on its pad at Cape Canaveral. But the U.S. had learned that the development and debugging of such complex weapons was a slow and painful process, filled with spectacular failures.

If the U.S. was making progress, the Soviet Union was not standing still, as shown by Khrushchev's speech on November 14 at a reception during the All-Union Congress of Soviet Journalists: "A few years ago I said in a speech that an intercontinental ballistic missile had been developed in our country. Then, many public leaders in capitalist countries stated that Khrushchev was probably just boasting. When we started production of these rockets, I said that in our country intercontinental rockets were on the assembly line. Again they began to say this could not be, that Khrushchev was boasting again." At this point, Deputy Premier Anastas Mikoyan interrupted to add: "Let them make such a boast themselves."

Khrushchev continued: "You can boast, but you must boast in such a way that all the world should see what you are boasting about. When we made such a boast all the world saw how our rocket soared to the moon and landed there. So this is no empty boast; these are real facts. I think, dear comrades and members of the Presidium, that I will reveal secrets, and at the same time I want to be understood correctly: we do not want to frighten anyone but we can tell the truth—now we have such a stock of rockets, such an amount of atomic and hydrogen warheads, that if they attack us we could wipe our potential enemies off the face of the earth. . . . It is a dreadful weapon for those who would like to unleash a war. . . . By the way, I shall reveal—and let the people abroad know it, I am making no secret of this—*that in one year 250 rockets with hydrogen warheads came off the assembly line in the factory we visited. This represents millions of tons in terms of conventional explosives. You can well imagine that if this lethal weapon is exploded over some country there will be nothing left there at all* [emphasis added]."

The Soviet leader said that Russia was "ready to sink all of this in the sea in the interest of ensuring peace on the earth . . . We are ready to destroy all these weapons of ours at once, if other countries will follow our example." Khrushchev concluded: "But things are going well with us. Our economy is developing, the might of the Soviet Union is rising. Such is our life, comrades! Not a bad life, comrades?" Foreign observers could not recall when they had heard the Russian leader speak with greater cockiness about Soviet power.

In Dr. Henry A. Kissinger's book *The Necessity For Choice*, published in 1960, the foreign affairs specialist wrote: "There is no dispute that the 'missile gap' will materialize in the period 1960–64. The only controversy concerns its significance. It may mean that we could lose if the Soviet Union struck first. In that case we would be fortunate if we escaped a surprise attack."

6

THE SLIT IN THE IRON CURTAIN CLOSES

It was 1960, the start of the projected period of maximum peril when many feared that Soviet superiority in ballistic missiles could neutralize the Strategic Air Command's bomber force. At best this would surely encourage an aggressive, expansionist Soviet foreign policy; at worst there would be risk of surprise thermonuclear attack.

It also was the year of a Presidential election in which the Missile Gap, and its associated Space Gap, were certain to be important issues. During the first week in January, President Eisenhower's State of the Union message sought to reassure the uneasy nation: "Our long-range striking power, unmatched today in manned bombers, has taken on new strength as the Atlas intercontinental ballistic missile has entered the operational inventory." The President said that "in fourteen recent test launchings at ranges of over 5,000 miles, Atlas has been striking on an average within two miles of the target . . ." He also noted that the current fiscal 1960 year would show a modest surplus and that a balanced budget would be submitted for fiscal 1961.

The Soviet response came a few days later, when Premier Khrushchev addressed the Supreme Soviet on January 14:

"The United States has set itself the task of catching up with the Soviet Union in the production of rockets in five years. They will naturally make every effort to raise their rocketry from the state it is now in and reach a better position, but it would be naive to think that we are meanwhile going to sit with arms folded." Then, in a move that bespoke confidence in the ballistic missile as the "ultimate weapon" and of the Soviet Union's superiority in the new technology, Khrushchev announced plans to reduce Russian armed forces by one-third!

Six days later, the Russians fired a new long-range missile, so powerful that it could not be tested for full range even within the vast reaches of the Soviet Union. The Russians cautioned international shipping to stay outside the target area, approximately a thousand miles southwest of Hawaii. The missile, according to the Russians, impacted within 1.24 miles of its aiming point. Eleven days later, on January 31, the Russians repeated the feat. The Soviets said that the large multistage rocket, which had covered nearly 8,000 miles, was intended to launch new and larger spacecraft, as was done subsequently. But it was obvious that the same 8,000-mile missile could deliver an H-bomb to the most remote part of the U.S.

The $41-billion fiscal 1961 Defense Department budget submitted to the Congress was no larger than the previous year's figure, but the $3.8-billion item for missiles represented a $600-million increase over the preceding year. Defense Secretary Thomas S. Gates, Jr., testifying before the House Defense Appropriations Subcommittee, reported that new intelligence estimates of Soviet ICBM strength had resulted in a downward shift in the 3:1 edge that only a year earlier had been projected for the 1960–63 period. The new intelligence estimates indicated that the Russians would enjoy only a "moderate numerical superiority" during the next three years, Gates stated. He said that the time of maximum Soviet advantage in numbers of ballistic missiles relative to the U.S. inventory "appears to peak during the 1962 period." Senator Symington, long an administration critic on the Missile Gap, promptly charged that national intelligence estimates "have been juggled so the budget books may be balanced."

The following week, during testimony before the House

Science and Astronautics Committee, Gates attempted to clarify the basis for the revised intelligence figures. "During the past year we have continued to acquire information which has enabled U.S. intelligence, *for the first time,* to estimate the probable Soviet ICBM inventory and its expected build-up with time. We also have an informed estimate of the performance of the Soviet ICBM . . . heretofore we have been giving you intelligence figures that dealt with theoretical Soviet capability [emphasis added]." Previous intelligence figures had been based on estimates of how many ICBMs the Soviet Union was thought to be capable of manufacturing. Now, thanks to U-2 reconnaissance flights over Russia, the U.S. had some "hard" evidence on which to base its new estimates. But Gates could not be so explicit because the U-2 program was still a closely guarded secret. It would not long remain so.

Symington remained unconvinced, and challenged the administration's overall appraisal even on the basis of the new intelligence figures. "The truth is," Symington charged, "that if we compare the ready-to-launch missiles attributed to the Soviets on the new intelligence basis with the official readiness program for U.S. ICBMs, the ratio for a considerable length of time will be more than three-to-one." Senator Lyndon Johnson, chairman of the Senate Space and Preparedness committees, also challenged the administration's position: "Recent official statements concerning our defense program for 1961 and beyond have been rosy and reassuring. It is startling and disturbing to find, however, that these reassurances depend upon changing the yardstick for measuring the Soviet threat."

Both Johnson and Symington had aspirations to be the Presidential candidate of the Democratic party, whose nominating convention was only a few months away. Symington accidentally had learned only two months earlier, during his visit to the base at Adana, Turkey, that U-2s were involved in a top-secret reconnaissance effort, but he had not been officially briefed on the results of the reconnaissance missions. Even among those who did have access to the results of the U-2 missions there was disagreement over what conclusions should be drawn from the lack of photographic evidence of widespread Soviet ICBM deployment. This was especially true because the limited U-2 missions had not been able to obtain photos of the entire Soviet land-mass.

The continuing Soviet achievements in space, while not of immediate military significance, served to underscore the USSR's pronounced lead in powerful rockets which were of military import. The impact of Soviet space spectaculars on U.S. allies and the uncommitted nations was discussed candidly by the director of the U.S. Information Agency, George V. Allen, in testimony before the House Science and Astronautics Committee during the last week in January:

> The achievement of placing in orbit the first earth satellite, without great advance fanfare, increased the prestige of the Soviet Union tremendously. Then came the two dramatic and successful Soviet moon shots followed by the failure of our own. As a consequence of these events, the seesaw seems to have tipped solidly in the Soviet direction in world opinion. It is hardly an overstatement to say that space has become for many people the primary symbol of world leadership in all areas of science and technology . . . The principal danger in the situation seems to me to be the cockiness which these successes have engendered in Soviet officials themselves. If it were a question merely of competition in scientific achievements, no one could properly begrudge the Soviets their magnificent successes . . . *However, if this new found Soviet cockiness (arrogance is not too strong a word) translates itself into adventuresomeness in foreign affairs, the world is in for a good deal of trouble* [emphasis supplied].

On February 21, as President Eisenhower prepared to depart for a tour of South America, he spoke to the nation on radio and television, ostensibly to discuss his upcoming goodwill visit. But it soon became apparent that the President had other things on his mind. Eisenhower sought to reassure the public by speaking of "our many hundreds of Air Force bombers deployed the world over—each capable of unleashing frightful destruction—[which] constitute a force far superior to any other, in numbers, in quality and in strategic location of bases." The President also noted that the nation's first Atlas missile squadron had recently become operational and that the first Polaris-armed submarines would soon go on station.

While the controversy increased over the latest intelligence estimates and the Missile and Space Gaps, the nation's ballistic-missile program continued to demonstrate evidence of progress. On January 6, 1960, a second Atlas ICBM had successfully flown the full 6,325-mile range and impacted on the intended target area in the South Atlantic. On February 2, after nine months of failures, the Titan had achieved successful ignition of both of its stages and had flown a 2,200-mile trajectory. Then, on February 24, another Titan was successfully fired from Cape Canaveral and flew a 5,500-mile trajectory, but a third Titan launch during the first week in March failed, because of a second-stage malfunction.

Despite the President's confident assurances in late February, by mid-March the administration was seriously considering a request for additional funds to step up production of the Atlas and Polaris missiles, as Defense Secretary Gates acknowledged in testimony before a joint session of two Senate committees. During the first week in April the President approved plans to seek funds for six more Polaris submarines, beyond the funds for three requested three months earlier and the 12 submarines already authorized. Each submarine would carry 16 Polaris missiles. The administration also sought funds to build 18 additional Atlas ICBMs, beyond the 40 in its original fiscal 1961 request. This action seemed to lend credence to the warnings of Congressional critics and to suggest that the administration was somewhat less confident of the U.S. strategic power than its official position.

The Polaris program was beginning to move along smartly. On March 29 the first missile to be fully guided from its own inertial guidance system had been successfully launched from the deck of the USS *Compass Island.* Sixteen days later, the first Polaris missile had been launched from below the ocean surface, using a tube to simulate the submarine, near San Clemente Island, Calif.

• • •

Plans for a four-power summit meeting to be held in Paris beginning May 16, 1960, had been announced in mid-December of 1959. It was a foregone conclusion that Premier Khrushchev would try to capitalize on the Missile Gap and the Soviet Union's widely accepted preeminence in this "ultimate weapon." To prepare the President with an up-to-date appraisal of Soviet missile strength, a U-2 reconnaissance mission had been flown from an air base in Pakistan on April 9. This was the first U-2 mission over southern Russia in some months. When these U-2 photos were developed, analysts reportedly discovered evidence of new construction efforts that looked suspiciously like ICBM sites. So plans were made for a follow-up flight a month later, when the construction would be more advanced and its intended purpose might be more clear.

The second U-2 flight departed on May 1 at 6:26 A.M. local time. Its pilot was Francis Gary Powers. His destination was Bodo, Norway, and his flight path would take him over Tyuratam, the Soviet Union's largest missile-test facility. Barely four hours after take-off, the Black Lady of Espionage, as the U-2 had been called, was a tangled wreckage near Sverdlovsk in central Russia and Powers himself was a prisoner.

Aside from the international repercussions that resulted, including the collapse of the Paris conference the following week, the incident came at an especially unfortunate time. For decades, even during World War II when the U.S. and the USSR were allies, the Russians had been able to maintain almost airtight security about their military strength, except when it suited their purpose. Only within the past four years, with the use of the U-2, had the U.S. been able to partially penetrate the Iron Curtain. Thanks to the U-2, we had discovered that the new jet bombers which the Russians had unveiled in the mid-1950s were a "red herring," intended to mislead the U.S. while they pushed ICBM technology.

Now in the early 1960s, when the peril of the Missile Gap seemed the greatest, the U.S. had lost its primary source of hard intelligence information. (Eisenhower, in an effort to placate Khrushchev's ire during the first day of the Paris conference, had told the Russian leader that the U-2 flights over the USSR would not be resumed.)

Despite heavy censorship, testimony released from a Senate hearing on the U-2 incident provides a clue to the great value of the program's reconnaissance photos. According to Defense Secretary Gates, the U-2 flights had provided vital information on "airfields, aircraft, missiles, missile testing and training, special weapons storage, submarine production, atomic production and aircraft deployments . . ." The loss of the U-2 operations, Gates said, "has removed an important source of intelligence that has been a very successful program over the past."

Soviet officials admitted that they had known about the U-2 over-flights almost since their inception, but that the airplane's unique design, which enabled it to fly at an altitude of roughly 80,000 feet, had kept it out of reach of existing Russian interceptors and anti-aircraft missiles. Now Premier Khrushchev boasted that thanks to a "remarkable new anti-aircraft missile" such aircraft could no longer be "sent into our skies with impunity."

The Soviet leader played down the military value of the U-2 missions. During a visit to Bucharest in June, while addressing the Rumanian Communist Party Congress, Khrushchev said: "I assert that the data obtained by the spy flights are of no importance . . . We know that the spy flights were carried out over regions which have no rocket bases. We know that two or three years ago, regions where experimental rocket launching sites were situated were photographed. Only the experimental grounds for launching rockets were photographed, not the rocket bases of military and strategic importance . . ." Was Khrushchev correct? Or was this a case of "whistling past the graveyard"? Only the Soviets could be certain. Only they could know with absolute certainty how many, or how few, Russian ballistic missiles were operationally deployed.

On June 28 the USSR announced plans for another series of long-range missile tests which would impact in the Pacific and it warned ships and aircraft to stay clear of the area, a thousand miles southwest of Hawaii. The Russian announce-

ment said one of the new giant rockets had been used on
May 15 to orbit the 10,000-pound Sputnik-5, the largest satellite by far ever launched. The achievement had been upstaged
in the aftermath of the U-2 incident, as had the U.S. launch
on May 20 of an Atlas which had flown more than 9,000 miles
and impacted in the Indian Ocean.

On July 1, Russian fighter planes operating out of eastern
Siberia shot down a U.S. RB-47 reconnaissance aircraft over
the Barents Sea between Alaska and Siberia. The U.S. said
the aircraft, which had been probing Soviet air defense radars
as the Russians often did with our radars in Alaska, had been
shot down while it was 30 miles outside Soviet territory. The
Cold War was getting hotter.

July 20, 1960, was a red-letter day for the Navy Polaris-
missile program. The first attempted launch of a Polaris from
a submerged submarine, the USS *George Washington,* successfully flew a thousand-mile trajectory. This was followed
within a matter of hours by a second successful Polaris launch
from the same submarine.

As the Republicans prepared to meet in Chicago during
the last week in July for their national convention, Premier
Khrushchev issued a stern statement about any U.S. "aggression" against Cuba and Fidel Castro. The Soviet Union, he
warned, would retaliate with its long-range ballistic missiles.
The Republicans already had decided to strengthen the defense plank in their 1960 platform, partially at the urging of
New York's Governor Nelson A. Rockefeller. The revised
platform said: "Swift technological changes and the warning
signs of Soviet aggressiveness make it clear that intensified
and courageous efforts are necessary for the new problems of
the 1960s . . ." During the same week, the Soviet Union vetoed
a U.S. resolution that the United Nations Security Council
conduct an impartial investigation of the shooting down of
the RB-47.

The following week, after the Democratic party national
convention had selected Senator John F. Kennedy and Senator
Lyndon B. Johnson as its candidates, both of whom had been
critics of the administration's defense policies, President
Eisenhower approved the spending of $476 million of an extra

$621 million which Congress had earlier appropriated beyond the funds requested for defense by the administration. Until this Presidential action, the authorized funds had been impounded. The President's message to Congress, explaining his decision to spend the added funds to accelerate key defense programs, noted that "we have seen an intensification of Communist truculence. Indeed, the Soviet dictator has talked loosely and irresponsibly about a possible missile attack on the United States. An American aircraft has been attacked over international waters . . . The Soviet delegation has walked out of the Geneva disarmament negotiations . . ."

On August 15, the first exclusively operational ICBM base, near Cheyenne, Wyo., was turned over to the Strategic Air Command. (The earlier one at Vandenberg AFB was also used for training and military satellite launches.) Three out of the 24 Atlas missiles planned for Warren AFB, Wyo., were declared operational.

Presidential nominee Kennedy served notice that he would "be vigorously criticizing various aspects of current national security policies, but that criticism will be responsible and constructive." Speaking on August 26 in Detroit to the Veterans of Foreign Wars, Kennedy said "the missile lag looms larger and larger ahead." The same week that Kennedy spoke, an operational version of the Titan (Titan-J) made a successful 5,000-mile flight from Cape Canaveral, the second for this configuration. But this and recent Polaris successes were tempered by the disappointing results of a Strategic Air Command exercise to demonstrate the readiness of the first operational Atlas site at Vandenberg AFB. When SAC crews tried to launch an Atlas, the effort had to be aborted. A second Atlas was launched, but it had to be destroyed shortly after take-off when it strayed from its intended course. The third Atlas, launched in mid-August, got off successfully but landed far wide of its intended target area in the Pacific.

A few weeks later, the Navy also took its lumps. Plans called for the launch of two Polaris missiles in quick succession, within six minutes, from a submerged submarine, the USS *Patrick Henry*, off the Florida coast. The first Polaris cleared the water (propelled from its tube by compressed air), but its engine failed to ignite and it ignominiously plopped back into the ocean. Another attempt was made several hours later, and this time the Polaris engine ignited, but the missile went out of control and tumbled into the ocean.

Senator Kennedy stepped up his attack on Eisenhower defense policies. Speaking in Seattle on September 7, he called for accelerated production of the Polaris, Minuteman and other missiles. Five days later, in Seattle, Kennedy charged that the risk of surprise attack by the Russians would grow as their missile lead increased. Republican Presidential nominee Richard M. Nixon no longer attempted a blanket defense of the military policies of the administration of which he had been a part. Speaking in California during the closing weeks of the campaign, Nixon said: "We must anticipate that our defense expenditures will go up rather than down in the period immediately ahead of us . . . We cannot rely exclusively on an existing weapon, as the last Administration relied on the manned bomber, when new weapons such as ICBMs threaten to make them obsolete."

Less than a month after the USS *Patrick Henry* had failed on its two successive attempts to launch a Polaris, it returned for another try. This time, during the period between October 13 and 18, the *Patrick Henry* got off four successive missile flights. All were launched while the submarine was submerged some 500 miles off the Florida coast. On the strength of this, the Navy announced in late October that its first Polaris submarine, the USS *George Washington,* with 16 missiles aboard, would become operational on November 15. And the USS *Patrick Henry* would join the fleet before the end of the year. The Navy also disclosed plans to begin tests soon on a longer-range version of the Polaris called the A-2. It would have a range of 1,500 miles, compared with 1,200 miles for the original A-1. Although the new model was slightly longer, it could be fired from the same submarines.

On November 5 the Air Force announced that it had advanced the initial operational date for its new Minuteman by one year. It now hoped to have the first squadron of 50 Minuteman missiles bedded down in their underground silos by July of 1962. The Air Force had not yet fired a complete three-stage Minuteman, but tests on missile components and simulated silo launches had been encouraging.

Five days later the Navy successfully fired its new longer-range A-2 Polaris missile, which flew a 1,600-mile trajectory from Cape Canaveral. The longer-range A-2 would enable the Polaris submarines to strike deeper into enemy territory, if need be, or to launch while the submarine was a greater distance away and therefore less vulnerable to enemy anti-submarine warfare defenses. On December 5 a second A-2 missile was successfully fired. Eleven days later, a SAC crew at Vandenberg AFB got off a successful Atlas launch, ending the previous string of three successive failures. The missile's reentry vehicle impacted 4,400 miles away in the target area in the Pacific, near Eniwetok Atoll.

As 1960 drew to a close and the new Kennedy administration prepared to take office, there was reason for modest optimism that the frightening Missile Gap had begun to close at long last. The year to come would provide some surprising disclosures about the Missile Gap.

7

THE BERLIN CRISIS,
SUPERBOMBS AND A NEW ERA

The outgoing President's final State of the Union message on January 12, 1961, responded briefly to the preelection criticism of the Eisenhower administration's defense policies which had helped give Kennedy his razor-thin victory. Eisenhower said that "the 'bomber gap' of several years ago was always a fiction and the 'missile gap' shows every sign of being the same." But four days later, when the outgoing President submitted the fiscal 1962 budget, which included a modest increase in defense spending, he appeared less sanguine in the accompanying budget message: "The advent of nuclear armed intercontinental ballistic missiles in the hands of a potential adversary has confronted this nation with a problem entirely new in its experience. The speed with which these weapons can be delivered against us and their tremendous destructive power make them suited to use by an enemy for surprise attack."

Premier Khrushchev had promptly dispatched a warm telegram of congratulations to Kennedy immediately after his election, encouraging the President-elect to hope for improved relations between the two powers. Kennedy responded with several conciliatory expressions in his January 20 Inaugural

Address. "Let us never negotiate out of fear. But let us never fear to negotiate," Kennedy said. Within the week, the Russians responded by announcing that they would release two surviving airmen from the RB-47 that had been shot down over the Bering Straits. Two days later the Kennedy administration invoked stiff controls on speeches delivered by all defense officials, civilian and military, to avoid provocative Get-Tough-type public statements.

The new President also ordered a fresh examination of the U.S. position on a nuclear test ban treaty in the hope of coming up with proposals that might break the long-deadlocked negotiations at Geneva, which had been recessed earlier for this purpose.

But Kennedy did not intend to rely prematurely on the new warmth that seemed to be radiating from the Kremlin. In his January 30 State of the Union message the President disclosed that he had requested his new Defense Secretary, Robert S. McNamara, to reappraise the nation's military capabilities and strategies and to submit a preliminary report by the end of February. On the basis of this analysis, Kennedy said he would take appropriate action. Meanwhile, he announced plans for a modest acceleration of the Polaris program.

On February 4 the Russians provided another reminder of their preeminence in large rocket boosters by launching the largest satellite ever placed in orbit, weighing 14,292 pounds. Eight days later they repeated the feat and then launched a smaller spacecraft from the satellite toward Venus, the first planetary probe ever attempted.

The day before the nuclear test ban treaty talks resumed in Geneva on March 21, the *Times* reported that the U.S. was prepared to offer concessions on some controversial issues. But the March 22 issue reported that the opening speech by Soviet delegate Semyon K. Tsarapkin "went a long way toward confirming the Western Powers' worst fears that the Russians had lost interest in working out a mutually acceptable treaty."

The following week, as a result of McNamara's new appraisal of U.S. defenses, President Kennedy requested $2 billion more than had been sought in the Eisenhower budget for fiscal 1962. The added funds would be used to further accelerate and expand the Polaris program, to expand Minuteman production facilities and to make the SAC bomber fleet more secure against surprise missile attack. This would permit 12 percent of the B-52 fleet to be placed on continuous airborne alert and, if necessary, increase the number of B-52s and B-47s on 15-minute ground alert to nearly half of the total SAC fleet.

In Geneva, the Soviet delegate took a more accommodating attitude on April 10, indicating Russian acceptance of the Western proposal for a control commission to police a test ban, after the U.S. and Britain had agreed to accept the Soviet demand for weapons parity. Two days later, on April 12, the Soviets achieved still another space first by launching a human cosmonaut, Yuri Gagarin, into orbit. The event, according to the *Times* editorial writer, "presents striking evidence that the Soviet Union possesses much more powerful rocket engines than the United States for the launching of intercontinental ballistic missiles."

Before another week had passed, the abortive Bay of Pigs invasion of Cuba was under way. Moscow promptly (and correctly) blamed the U.S. for the attack, despite initial denials by the U.S. government. Khrushchev announced that the Soviet Union would help Cuba in its hour of need, but it soon became apparent that the Fidel Castro government required no outside assistance. There was a curious omission in Khrushchev's blustery offer to assist Cuba. He carefully avoided any threat to launch long-range missiles against the U.S. Yet only nine months earlier, on July 9, 1960, he had publicly threatened to use missiles to protect Cuba: ". . . the United States is no longer at an unreachable distance from the Soviet Union," the Russian leader had said then. "Figuratively speaking, in case of necessity, Soviet artillery-men can support the Cuban people with their rocket fire if the aggressive forces in the Pentagon dare to start intervention against Cuba."

Within several days after the invasion had been put down, Khrushchev surprised the West by "turning the other cheek." He announced on April 22 that despite the Cuban adventure the Soviet Union would continue to seek a broad agreement with the U.S. on outstanding issues. Two weeks later, during a visit to the capital of Soviet Armenia, the Russian leader predicted there might soon be disarmament discussions between the two countries.

On May 16 the White House announced that the President would meet with Khrushchev in Vienna the following month; three days later the date was fixed as June 3–4. It appeared that the Soviet leader was anxious for improved relations, despite Cuba, despite continuing difficulties over Communist pressures in Laos, Thailand and South Vietnam, despite the very recent Kennedy administration announcement of plans to provide military aid to Thailand and South Vietnam, and despite the ever-present issue of West Berlin. If Khrushchev was as eager for détente as he seemed to be, Kennedy might return from Vienna with enough to erase the sour taste of the Bay of Pigs and to repair his own image which had suffered as a direct result.

The U.S. appeared to be in a somewhat stronger bargaining position in terms of the Missile Gap than a year earlier when Eisenhower prepared for the Paris Summit Conference. The U.S. now had three Polaris submarines deployed, each armed with 16 missiles, and two more were slated to join the fleet by September to raise the total Polaris arsenal to 80 weapons. In addition to the several squadrons of IRBMs based in Britain, Italy and Turkey, the U.S. now had approximately two dozen Atlas ICBMs deployed, and the first successful launch of the more advanced Titan from a hardened 146-foot-deep underground silo had been achieved on May 3. The first successful launch of the new Minuteman missile from Cape Canaveral had occurred on February 1, when the new-generation ICBM had flown a 4,600-mile trajectory.

The giant BMEWS radar in Thule, designed to provide 15 minutes' warning of an enemy ICBM attack to enable SAC bombers on ground alert to become airborne, was now operational, and a second BMEWS radar in Alaska, which could detect missiles coming from Siberia, would go into service in July. This, coupled with the recent measures to place one-eighth of the SAC B-52 fleet on airborne alert and to increase the number of bombers on 15-minute ground alert, had greatly diminished the possibility of a surprise missile attack that could decimate the SAC fleet.

The days before the Vienna meeting brought mixed portents. U.S. Ambassador Llewellyn Thompson, after conversations with Khrushchev, reported back that the Russian leader was in a cocky and aggressive mood. But on May 31, during a stopover in Bratislava, Czechoslovakia, en route to Vienna, Khrushchev spoke of the Soviet Union's desire "for relaxation of international tension."

An amiable spirit seemed to prevail during the first informal hours at Vienna. Khrushchev joked with Mrs. Kennedy, and press photos showed a smiling President and a beaming Khrushchev as the first day's talks were about to begin. At the end of the first day, a U.S. spokesman said the two leaders had had a "frank, courteous and wide-ranging exchange." But press photos taken at the conclusion of the second, and final, day's talks showed a grim Khrushchev and a somber Kennedy. *The New York Times* reported that the conference had "ended in hard controversy." On June 6, after a brief stop in London for talks with the British Prime Minister, the President addressed the nation on television from Washington. "I will tell you now that it was a very sober two days," the President reported to the American public.

Nine days later the full dimensions of the "sober two days" were revealed when the Soviet leader announced in Moscow that the Russians had set a December 31 deadline for signing a German peace treaty. If necessary the Soviets would sign a unilateral peace treaty with East Germany, Khrushchev said. This would precipitate a grave crisis over the status of, and access to, West Berlin. The lead editorial in the June 16 issue of the *Times* called Khrushchev's remarks "a chilling speech." It noted that the Russian ultimatum was essentially the same one that the Russian leader had presented to Kennedy at Vienna, and added: "It is easy to understand why the President used the word 'sober' to describe his meeting with the Premier."

The uncompromising attitude of the Soviet negotiator at the Geneva nuclear test ban treaty talks increased U.S. fears that the Russians had no desire to reach agreement and might secretly be preparing to resume weapons tests. This prompted a U.S. warning that it could not forgo testing indefinitely if negotiations remained deadlocked. Khrushchev responded on June 21, declaring that Russia would initiate nuclear tests immediately if the U.S. abrogated the fragile test moratorium. His speech "bristled with threats of nuclear destruction 'against the imperialist aggressors,'" the *Times* reported on June 22, and noted: "The speech displayed an uncompromising line on major policies and was delivered at an extraordinary meeting in the Great Kremlin Palace."

The war of nerves was accelerating. The next day the President held a three-hour meeting with his top aides on the Berlin crisis. Several days later he met with the National Security Council in a long session. Khrushchev announced on July 8 that he had canceled earlier plans to slash the size of the Soviet armed forces. He also disclosed that the Soviet defense budget would be increased by nearly one-third. The next day at Moscow's Tushino airport, the Russians unveiled a new swept-wing supersonic bomber as well as two new types of interceptor aircraft. It was the first public display of new Soviet military aircraft in five years.

The following day, the President ordered an "urgent review" to determine whether the U.S. should increase its military strength to meet the growing Soviet threat. On July 18 the U.S. formally warned the Soviet Union that it would defend its rights in Berlin. The next day, the President met with the Joint Chiefs of Staff. Within a week, on July 25, the President again addressed the nation to discuss the growing threat of war. In addition to announcing an expansion of the armed forces, he disclosed he would seek an additional $3.5 billion in defense funds. He spoke of expanded efforts in civil defense and of air raid shelters. But Kennedy declared the U.S. was willing to meet with the Soviets to "seek genuine understanding, not concession of our rights." The President's message found wide support both in the Congress and among the public at large.

On August 1 the Pentagon notified 64 Air Force Reserve units to be ready for a possible call to active duty. Several days later, Khrushchev spoke of the possibility of a conference to resolve the Berlin crisis, but warned that Soviet troops might mass along West European borders as a "precautionary measure."

Then on August 9 the Soviet leader warned that his country had the capability of constructing giant rockets which could carry a 100-megaton thermonuclear warhead (five thousand times the size of the bomb that leveled Hiroshima). During an impromptu talk at a Kremlin reception honoring the latest Soviet cosmonaut, Gherman Titov, Khrushchev said: "If you want to threaten us from a position of strength, we will show you our strength. You do not have 50 and 100 megaton bombs. We have stronger than 100 megatons. We placed Gagarin and Titov in space, and we can replace them with other loads that can be directed to any place on earth." The Soviet leader had made similar threats during recent discussions with John J. McCloy, the President's disarmament adviser, McCloy's aides subsequently revealed.

Despite the growing Berlin crisis and Khrushchev's hard talk, the U.S. disclosed plans to make one final effort to assess Soviet interest in reaching an agreement in Geneva on a test ban treaty by returning its chief negotiator, Arthur H. Dean. Three days later, on August 14, two Russian army divisions moved into position around Berlin to back up East German efforts to seal off its borders. The Soviet forces included tanks and artillery. On August 16 the U.S. Army alerted 113 Reserve and National Guard units for possible active duty. The following day it was announced that 1,500 additional U.S. troops were being sent to West Berlin and that Vice President Johnson would visit the beleaguered city.

On August 24 the White House issued a "solemn warning" to the USSR that any interference with Allied access to West Berlin would be "an aggressive act" for which the Soviet Union would bear full responsibility. The next day, the Pentagon ordered 76,500 reservists to active duty. The same day, East German Premier Walter Ulbricht softened the Communist position on Allied access to West Berlin. Two days later, Premier Amintore Fanfani of Italy, following his recent visit to Moscow, revealed that he had received a letter from Khrushchev in which the Russian leader expressed a willingness to negotiate on Berlin and other European problems.

But there was no sign of compromise at the negotiations

in Geneva. On August 28, Soviet delegate Tsarapkin brushed aside U.S. efforts to reopen negotiations by announcing that it was useless to even talk about a test ban treaty, except as part of an overall agreement on "general and complete disarmament." Three days later, on August 31, the explanation for Russia's attitude in Geneva became crystal-clear when the Soviet Union announced that it was unilaterally breaking the three-year moratorium and would resume nuclear-weapon testing. The Russian announcement said the USSR planned to test a series of superpowerful bombs with yields of 20 to 100 megatons. It explained that the Soviet decision was prompted by the "threatening attitude" of the U.S. over the Berlin crisis. The announcement also referred to Soviet rockets capable of delivering the giant new H-bombs "to any point on the globe."

Within twenty-four hours the USSR had exploded its first nuclear weapon in the atmosphere at Semipalatinsk in central Asia. There would be 14 more during the next 21 days and many more before the series was finally ended. In view of the extensive preparations normally required for such tests, it was clear that the Soviet decision to break the moratorium had been made many weeks earlier. The White House immediately denounced the Soviet action as "a form of atomic blackmail, designed to substitute terror for reason."

Newspaper reporters queried Pentagon nuclear experts on the destructive force of the 100-megaton bomb about which Khrushchev had boasted. They were told that a single weapon of this size would create *total destruction* over a 12-mile radius, or approximately *450 square miles,* and approximately half the people living in an area 12 to 20 miles away from the point of explosion would be killed. The radioactive fall-out from a single weapon could be expected over a 20,000-square-mile area, equivalent to the combined size of the states of Massachusetts, Connecticut, Rhode Island and New Jersey.

As the radioactive debris from the first Soviet atomic explosion was settling, Khrushchev was quoted as saying he had taken the action to shock the Western powers into negotiations on Germany and disarmament. Despite this attempt to deflect responsibility, the Soviet action created tremors in Belgrade, Yugoslavia, where leaders from twenty-four "uncommitted" nations were meeting. A shocked Indian Prime

Minister Jawaharlal Nehru warned that the Soviet action had brought the danger of war closer.

The next day, the U.S. announced it would send 72 jet fighters to Europe to participate in Allied war games. The following day, President Kennedy and British Prime Minister Harold Macmillan sent a joint message to Khrushchev, urging the Russians to refrain from nuclear tests in the atmosphere with their resultant fall-out. Two days later, on September 5, Moscow rejected the proposal, ridiculing it as "deceitful, unrealistic and propagandistic." The same day, Kennedy approved resumption of U.S. nuclear testing in underground sites but withheld approval for atmospheric tests.

Prime Minister Nehru and President Kwame Nkrumah of Ghana flew to Moscow on September 6 as spokesmen for the Belgrade conference, to personally appeal to Khrushchev to meet with Kennedy in an effort to avert World War III. The following day, Khrushchev granted a long interview to C. L. Sulzberger of *The New York Times*. The Soviet leader said he was prepared to meet with Kennedy, but once again spoke of Russia's 100-megaton bomb and its rockets which could deliver the warhead. The next day, Khrushchev publicly outlined his conditions for any negotiations on Berlin. He insisted that such talks must be held soon and that they must result in a German peace treaty. During the same day, the Western powers issued a stern warning to Moscow against any interference with Allied flights to West Berlin. The following day, Khrushchev responded to the Kennedy-Macmillan proposal by declaring that nuclear testing could be ended only by Western acceptance of Soviet proposals for a German peace treaty and complete disarmament. Later in the day the U.S. announced it would send 40,000 more troops to West Europe.

As if to add to the psychological impact of the superbomb explosions to come, Russia announced on September 10 that it would test a new series of more powerful rockets between September 13 and October 15. It warned ships and aircraft to avoid the target area in the central Pacific. Two days later the Russians exploded the first large thermonuclear weapon, with a yield of several megatons, near the Arctic island of Novaya Zemlya.

On September 13, as announced, the Russians fired one of

their new rockets on a 7,500-mile trajectory into the Pacific and subsequently reported that it had hit within one-half mile of its intended impact point. Soviet rocket chief Marshal Moska-lenko declared that Russian missiles were "invulnerable" and noted that they could strike any point around the world. The missile shot was orchestrated with two nuclear tests on the same day, one in the Arctic and the other in central Asia. Five days later the Soviets fired a second long-range ballistic missile into the Pacific, and Moscow reported it had impacted near the earlier shot. On the same day, the Russians tested a megaton-level weapon in the Arctic, the thirteenth in less than three weeks. The following day the Pentagon called 73,000 reservists to active duty.

One small encouraging sign on an otherwise clouded in-ternational horizon came on September 21 in New York City, when Secretary of State Dean Rusk met with Soviet Foreign Minister Andrei Gromyko, who had come, ostensibly, to at-tend the new session of the United Nations General Assembly. The next day, in Moscow, Khrushchev asserted that he was ready for negotiations at "any time, any place and at any level" to avert the risk of war over Berlin and Germany. Three days later a *Times* correspondent in Washington reported the "first signs" that both Moscow and Washington were begin-ning to acknowledge their eagerness for "honorable" negotia-tions.

President Kennedy addressed the UN General Assembly on the evening of September 25. He warned of the danger of nuclear war if the Soviet Union tried to ride roughshod over Western interests and commitments. But the President also suggested that a peaceful settlement was possible. At the UN, Kennedy seemed remarkably calm and reassured for the leader of a nation which appeared to be moving inexorably toward thermonuclear war—especially since its adversary seem-ingly enjoyed an overwhelming advantage in ICBMs which could destroy most of the American population centers within 30 minutes after Khrushchev gave the launch order.

An obscure clue to the reason for the President's confidence had appeared in *The Washington Post*, on the morning of the Kennedy address to the UN, in a column written by Joseph Alsop. Columnist Alsop not only had excellent Pentagon

sources but also was one of the President's circle of journalistic
intimates. The Alsop column was titled "Facts About the
Missile Balance." It began:

> Mixed but broadly encouraging results have been obtained
> by recent, exceedingly careful recalculation of the prob-
> able nuclear striking power of the Soviet Union. The
> most cheerful of these results is a sharp reduction in the
> number of intercontinental ballistic missiles with which
> the Soviets are credited.
>
> As is known, such figures always come in a spectrum
> form, beginning with a minimum and ending with a
> maximum. Where nuclear striking power is concerned, the
> maximum figure invariably originates with the Air Force
> for obvious reasons. And if the Air Staff agrees to a dras-
> tic cut in the maximum figure, as it has now done, one
> must tentatively accept this maximum as realistic and
> convincing.
>
> Prior to the recent recalculation, the maximum number
> of ICBMs that the Soviet were thought to have at this
> time was on the order of 200—just about enough to per-
> mit the Soviets to consider a surprise attack on the United
> States. *The maximum has now been drastically reduced,
> however, to less than a quarter of the former figure—well
> under 50 ICBMs, and therefore not nearly enough to allow
> the Soviets to consider a surprise attack on this country.*
> [Emphasis supplied.]

Alsop went on to say that the Soviet fleet of long-range
bombers was unchanged, numbering about 150, or about one-
quarter the size of the U.S. fleet of B-52s. Further, that the
USSR was now credited with having about 300 medium-range
ballistic missiles capable of reaching West European countries.

The Alsop column concluded: *"For the present, what is
chiefly important is to make Khrushchev understand that if he
pushes the Berlin crisis to the ultimate crunch, the great
power that the United States still possesses will not remain
unused* [emphasis added]."

To the casual reader, unaware of how a skillfully planted
"leak" is often used by top government officials in Washington,

it might have seemed that the Pentagon had recently discovered some arithmetic error in its previous tabulations of estimates of Soviet missile strength. To anyone familiar with the long, bitter arguments between the USAF and CIA over the latter's far lower estimates of Russian missile strength, as Alsop was, the fact that the new figures had been endorsed by the USAF's Air Staff was a significant tip-off that the new appraisal was based on something far less contestable than "guestimates." It was!

Geopolitics has always been a high-stakes poker game in which bluff has played a key role. For more than fifteen years, the U.S. had been engaged in such a game with the Soviet Union. The USSR's strategists had been able to see most of the American cards because of our open society and free press, while Russia had been able to hold its own cards close to its chest. Briefly, during the late 1950s, the U-2 had been able to penetrate the Iron Curtain sporadically to reveal, for example, that the Russians were not building a large fleet of jet bombers as their public displays had suggested.

With the shooting down of the Powers U-2 in May of 1960, and the U.S. decision to renounce further U-2 flights over the USSR, the Russians had regained their precious advantage of secrecy. Khrushchev, master international poker player that he was, had taken full advantage of the resulting gap in U.S. intelligence. But now, in September of 1961, it had become an entirely new poker game. The U.S. had a photographic inventory of Soviet missile sites, located near the Trans-Siberian railroad that was needed to transport the giant missiles to their launch pads. *The photos had been provided by American reconnaissance satellites.*

The U.S. now knew that the Russians had built only a handful of their first-generation ICBMs, not the hundreds that had been estimated by American intelligence agencies only a couple of years earlier. This meant that the U.S., with its large fleet of long-range SAC bombers and its growing fleet of Polaris missiles and Atlas ICBMs, really had an *overwhelming* advantage in strategic weapons, rather than being on the short end. For the first time, the U.S. as well as the USSR had an accurate count of the strategic power of its adversary.

This, Kennedy and a few of his top advisers knew on the evening of September 25 as he spoke to the UN. It explains why he could so confidently take off from New York for a week's vacation at Newport, R.I. Five days later, Rusk and Gromyko held their third talk in New York, this one lasting some four hours. A week later, on October 6, President Kennedy returned to Washington for a two-hour meeting with the Soviet Foreign Minister. The press reported that the talks had gone badly, yet the President confidently returned to Newport to relax with his family aboard the Presidential yacht.

The successful firing of a fifth Soviet long-range missile into the Pacific was announced on October 13, with emphasis on its "fundamentally new guidance system" to explain the missile's high accuracy. The next day, the U.S. and Canada conducted the largest air defense exercise ever attempted. All civil aircraft in both countries were grounded for the 12-hour drill.

On October 16 the U.S. announced that another Atlas missile base with nine ICBMs, located near Topeka, Kan., had become operational, raising the total Atlas missiles in readiness to 48. The Topeka base, like another recently completed near Spokane, Wash., provided some protection for the missiles against enemy warheads exploding more than a mile away. *The New York Times* story on the new Atlas base, written by Richard Witkin, said the U.S. now estimated that the Soviets had 35 to 75 operational ICBMs, but gave no source for these figures.

The next day, October 17, the 22nd Communist Party Congress opened in Moscow, with top Communist leaders from around the world in attendance, including Red Chinese Premier Chou En-lai. During Premier Khrushchev's six-and-a-half hour speech, he disclosed that Russia planned to end its series of nuclear tests late that month by exploding its much-vaunted 50-megaton superbomb. *But Khrushchev's real blockbuster came when he announced that he had decided to cancel his December 31 deadline for signing a German peace treaty*—provided the West showed its readiness to negotiate on the German question. The Soviet leader said: "We have the impression that the Western Powers were displaying a certain understanding of the situation and that they were inclined to seek a solution for the German problem . . . on a mutually acceptable basis."

Three days later, six countries which would be in the direct path of fall-out from the Russian superbomb tests—Denmark, Norway, Sweden, Japan, Canada and Iceland—asked the UN General Assembly to issue a "solemn appeal" to the Russians to cancel their test. This came a day after the U.S. had announced that it would resume atmospheric testing unless a test ban treaty was promptly negotiated.

Five days after Khrushchev had withdrawn the German peace treaty deadline, the administration decided that the American public should be told, if only cryptically, what the Russians now knew. On October 22, Deputy Secretary of Defense Roswell L. Gilpatric spoke to the Business Council, a group of distinguished businessmen, meeting in Hot Springs, Va. When the Pentagon official met with reporters, he emphasized that his speech had been cleared "at the highest level."

Gilpatric said that Russia's "Iron Curtain is not so impenetrable as to force us to accept at face value the Kremlin boasts." He warned that the U.S. "has a nuclear retaliatory force of such lethal power that an enemy move which brought it into play would be an act of self-destruction on his part." Gilpatric cited SAC's fleet of 600 long-range B-52 bombers, even more of the shorter-range B-47s, the Navy's six Polaris submarines with a total of 96 missiles and the several dozen Atlas missiles then operational. (One of the six submarines was not yet operational but was undergoing sea tests.)

Gilpatric's remarks were featured on the front page of *The New York Times* and *The Washington Post,* despite the obscure location at which he spoke. The *Post* went on to say that the U.S. position in negotiations with the Soviet Union "has taken a stiffening turn in the past 10 days," thus contradicting Khrushchev's Party Congress statement. The newspaper called Gilpatric's speech "the toughest made to date by a high Kennedy Administration official." The day after Gilpatric spoke, Secretary of State Rusk appeared on the American Broadcasting Company's television interview program *Issues and Answers.* Rusk volunteered that he had helped review Gilpatric's "well-considered" statement and added: "We are not dealing . . . these days from a position of weakness." When reporters pressed Rusk to explain whether the new administration view on the relative strengths of the U.S. and USSR resulted from an increase in American strategic power or a downgrading in estimates of Soviet strength, Rusk dodged the issue.

The very next day, the Soviets produced the largest man-made explosion in history—not the promised 50-megaton bomb but a hefty 30-megaton weapon—in the Arctic. The same day, Soviet Minister of Defense Rodion Y. Malinovsky told the Party Congress in Moscow that the Russians had created a new Strategic Rocket Forces which included "1,800 excellent units" capable of striking "any point on the globe without a miss." Marshal Malinovsky also said: "I must report to you especially that the problem of destroying missiles in flight has been successfully solved."[*]

It was curious that the Soviet official should now claim that Russia had developed a successful anti-ballistic missile (ABM) because Soviet leaders had boasted in the past that the ICBM was invincible. (At the time, the consensus of Western scientists was that no fully effective ABM defense was possible.) Was this a subtle admission that the Russians now recognized that they no longer dominated in the field of ICBMs and were trying to downgrade the import of America's growing missile arsenal?

On October 28, Khrushchev announced that Russia would proceed with plans to explode its 50-megaton superbomb, despite "hysterical protests," i.e. the UN General Assembly's "solemn appeal." He complained that the protesters did not understand the Soviet reasons for proceeding with tests—a veiled admission that the USSR no longer was on the favorable end of the Missile Gap. Two days later the Russians

---

[*] When U.S. officials failed to be impressed by Malinovsky's claim, Moscow Radio promptly took the West to task. It boasted that Russian rocket units "can destroy all means of delivery of atomic weapons in the air. In other words, there is an anti-missile for every single missile."

exploded a 58-megaton weapon in the Arctic, overshooting the target size by 16 percent.*

The superbomb produced no panic in the White House or in the Pentagon, nor in Western capitals. Even as the Berlin crisis had been escalating and the Soviets had been conducting their nuclear and missile tests for psychological advantage, the U.S. had been launching additional reconnaissance satellites. There had been one launched on July 7, another on August 30 and on September 12, and still another on October 13.

Despite U.S. secrecy, the Soviets knew of the spaceborne reconnaissance, if only from guarded statements in published Congressional hearings and articles in the American press. Perhaps, initially, the Soviets believed that the camera resolution obtainable from an altitude of more than 100 miles would be too poor to be of value. But by late 1961 Soviet officials had to face two harsh and unpleasant facts: first, U.S. officials now knew that it was the Soviet Union which was on the short side of the Missile Gap; and second, that America's advantage would soon grow to overwhelming proportions as new Minuteman ICBMs and Polaris missiles began to roll off the production lines to be deployed in underground silos and far-ranging nuclear submarines.

A new era had dawned: the age of strategic spaceborne reconnaissance—as significant as the advent of nuclear weapons. But there was no Hiroshima or Nagasaki to herald the event and so there were few who realized that a new era was at hand. Still fewer could then foresee the remarkable impact it would have on international relations in this century and far beyond.

---

* The announcement of the test of the 58-megaton weapon at the 22nd Party Congress was the first indication to the Soviet public that the Russians had resumed nuclear testing, even though they had broken the moratorium nearly two months earlier. So far as the Russian public knew, it was the U.S. that had first violated the moratorium by conducting underground nuclear tests, which the Soviet press had fully reported.

8

THE GENESIS OF
SATELLITE RECONNAISSANCE

The reconnaissance satellites that punctured the Iron Curtain in 1961 to expose the true status of the Missile Gap, and which could have an eternal effect on international affairs, had their genesis in a 324-page report published fifteen years earlier, on May 2, 1946. "This report," it began, "presents an engineering analysis of the possibilities of designing a man-made satellite . . . It is concluded that modern technology has advanced to a point where it now appears feasible to undertake the design of a satellite vehicle."

The report was the product of a group of technical specialists that had been formed a few months earlier at Douglas Aircraft Company in Southern California. The group was called Project RAND and by 1948 would be spun off and set up as an independent "think tank" under the name of The RAND Corporation. The new technology was the German V-2 missile, which had been developed largely from the pioneering work of early German rocket enthusiasts with visions of space travel. But once the German military had taken over sponsorship in the 1930s, there had been neither the resources nor the incentive for such projects, although some of the pioneers continued to dream of more peaceful applications.*

The RAND report explained: "If a vehicle can be ac- celerated to a speed of about 17,000 mph. and aimed properly, it will revolve on a great-circle path above the earth's atmo- sphere as a new satellite. The centrifugal force will just bal- ance the pull of gravity. Such a vehicle will make a complete circuit of the earth in approximately 1½ hours [for the 300-mile- high orbit assumed in the analysis]."

"Such a vehicle will undoubtedly prove to be of great military value. However, the present study was centered around a vehicle to be used in obtaining much desired scien- tific information on cosmic rays, gravitation, geophysics, terres- trial magnetism, meteorology and properties of the upper atmosphere. For this purpose, a payload of 500 pounds and 20 cubic feet was selected as a reasonable estimate of the re- quirements for scientific apparatus capable of obtaining results sufficiently far-reaching to make the undertaking worthwhile," the RAND report said.

"We can see no more clearly all the utility and implica- tions of spaceships than the Wright Brothers could see fleets of B-29s bombing Japan and air transports circling the globe. Though the crystal ball is cloudy, two things seem clear: (1) A satellite vehicle with appropriate instrumentation can be expected to be one of the most potent scientific tools of the Twentieth Century; (2) The achievement of a satellite craft by the United States would inflame the imagination of mankind, and would probably produce repercussions in the world comparable to the explosion of the atomic bomb." This prediction by scientist David T. Griggs would prove especially prescient barely a decade later when the first Rus- sian Sputnik went into orbit.

A chapter in the report by Dr. L. N. Ridenour discussing

---

* Wernher von Braun and two other Peenemünde scientists were arrested by the Gestapo on March 14, 1944, on the charge that they really were not interested in producing a useful weapon and only wanted to develop rockets for space travel. The scientists were released only after V-2 project chief General Walter Dornberger declared under oath that the V-2 program (which by this time had caught Hitler's fancy) could not continue without Von Braun and his two associates.

the potential military significance of satellites projected their use for meteorology to obtain an overview of cloud patterns and as communications relay stations, both important today for civil and military use. "It should also be remarked," Ridenour wrote, "that the satellite offers an observation aircraft which cannot be brought down by an enemy who has not mastered similar techniques . . . Perhaps the two most important classes of observation which can be made from a satellite are the spotting of points of impact of bombs . . . and the observation of weather conditions over enemy territory. Certainly the full military usefulness of this technique cannot be evaluated today. There are doubtless many important possibilities which will be revealed only as work on the project progresses." This would prove another prescient observation.

The preliminary RAND study suggested two possible rockets which would be capable of putting a 500-pound satellite into a 300-mile earth orbit. One was a four-stage rocket, estimated to weigh 234,000 pounds, which would be powered by the same fuels used in the V-2—alcohol and oxygen. The other rocket design, a more advanced concept involving only two stages, would use liquid hydrogen and oxygen as fuel. Its weight was estimated at 292,000 pounds, nearly ten times that of the V-2.

(The idea of multistage rockets for space flight is credited to Russian rocket pioneer Konstantin Tsiolkovsky, who suggested the concept at the turn of the century. A rocket is a relatively inefficient means for lifting weight. Fuel, tanks, valves, pumps and other parts of the engine make up most of the total vehicle weight, leaving little for useful payload. Ideally, every pound of weight ought to be jettisoned as soon as it has served its purpose. The practical approach to this ideal is to build a rocket in several stages, each of which is essentially a complete rocket, with each succeeding stage being smaller than the one below.)

In the four-stage rocket that RAND scientists envisioned, the first stage would power the vehicle to a speed of approximately 3,000 mph, at which time the first-stage fuel would be exhausted and this entire stage would be jettisoned. The second-stage rocket engine would then be ignited and would be used until a speed of approximately 7,600 mph had been achieved, at which point the second stage would drop off. The third stage would serve to boost the speed to approximately 12,000 mph before being jettisoned. Then the fourth stage, weighing about 2,900 pounds, including 500 pounds of useful payload, would be fired to boost the entire stage to the velocity of approximately 17,000 mph needed for earth orbit.

The RAND study, the first detailed technical analysis of an earth satellite executed in the U.S., focused all of its attention on the difficult problem of designing a launch vehicle. Almost no effort was devoted, understandably at that point, to the equally challenging task of how to design a useful satellite payload with the electronic technology that then existed. In 1946 all electronic equipment of the type to be carried in a satellite would have had to use vacuum tubes. Not only were these large and heavy, but they required large amounts of electric power. The heavy batteries required to supply this power would have further reduced the already frugal payload that could be carried.

By 1948, Bell Telephone Laboratories would invent the remarkable transistor, a device that could perform the function of a vacuum tube in a tiny fraction of its size and weight, and which consumed a hundredth as much electric power. By 1960, transistor (solid-state) technology would have advanced to the point where hundreds of individual transistors and associated components could be formed and electrically interconnected in a "microcircuit" no larger than a single transistor of the 1950 vintage. This solid-state technology also would produce silicon solar cells, capable of converting sunlight into electricity, which could provide electric power aboard certain types of satellites.

In addition to the formidable technical challenges which the RAND report cited, and the problem of achieving a useful satellite payload with the 1946-vintage electronic technology, the idea of an earth satellite program faced another serious obstacle—economics. RAND scientists estimated (optimistically) that it would cost $150 million and take five years to build and launch an earth satellite. In the early post-World War II period, when most Americans anticipated a

long and enduring period of peace, the U.S. defense budget
had been cut back drastically. It seemed foolish to spend
$150 million for so speculative and risky a project when more
obviously useful military projects suffered for lack of funds.
For example, in the spring of 1946 the USAF had been able
to provide only $1.4 million to Convair for its work in long-
range ballistic missiles, and by the following summer the
project had to be canceled because of the defense-budget
squeeze.

The USAF found sufficient funds to support several RAND
satellite studies. One "secret" report issued February 1, 1947,
was entitled *Aerodynamics, Gas Dynamics and Heat Transfer
Problems of a Satellite Rocket*. Another of the same date was
entitled *Study of Launching Sites for a Satellite Projectile*.

By 1950 the Cold War had emerged, tinged with hot spots
like West Berlin and Korea. The defense budget was much
larger, but there also were many military needs more pressing
than satellites. However, there was growing need to try to
find out what was happening behind the Iron Curtain. The
pace of RAND satellite studies was accelerated and resulted
in several more "secret" reports issued in April 1951. One was
titled *Utility of a Satellite Vehicle for Reconnaissance*; another, *Inquiry into the Feasibility of Weather Reconnaissance
from a Satellite Vehicle*.

By 1952 the U.S. had begun the development of an enlarged V-2-type missile, the Redstone, by Von Braun and
other former German scientists at the Army's arsenal in Huntsville, Ala. The Redstone, which would make its first successful
flight in 1953, was not powerful enough to launch a satellite,
but it was a start. The USAF was once again funding an
ICBM development at Convair, and by February of 1954 the
Teapot Committee's "secret" report would call for a crash
ICBM effort. If the Atlas program proceeded on schedule, by
1960 the U.S. would have the rocket booster needed to launch
a satellite, a spacecraft considerably larger than the one first
suggested by the 1946 RAND report.

Through tiny, occasional cracks in the Iron Curtain were
seen alarming signs of a Soviet build-up in strategic weapons,
and this whetted the appetite of U.S. defense planners for
better intelligence information on Russian activities. During

the 1952–53 period, RAND initiated a series of secret design studies on reconnaissance satellites under the code name of Project Feed-Back, indirectly sponsored by the CIA.

On March 1, 1954, a series of Project Feed-Back reports, classified "secret," was issued and circulated on a very restricted basis. One two-volume summary report bore the unclassified title *An Analysis of the Potential of an Unconventional Reconnaissance Method*. RAND sought to bolster its expertise in aerial reconnaissance technology, and in 1954 Amrom Katz, the chief physicist at the USAF's Aerial Reconnaissance Laboratory in Dayton, Ohio, left that post to join RAND. Katz had been employed in the Aerial Reconnaissance Laboratory at Wright-Patterson AF Base since 1940 and had played a key role in developing the cameras to be used in the Lockheed U-2.

By the mid-1950s those who could see great military and scientific possibilities in satellites were buoyed by the prospect of soon having large rockets, then being developed for the ICBM program, to launch spacecraft into orbit. But there was a loud discordant note. Defense Secretary Charles E. Wilson, who had come to the Pentagon from a long tenure at General Motors, was a practical down-to-earth man in every sense of that expression. Neither he nor President Eisenhower was quick to grasp the military potentialities of satellites.

The President had, on July 29, 1955, approved plans to launch a tiny satellite as part of the International Geophysical Year (IGY) program, to be called Project Vanguard, at the urging of scientists involved in the IGY effort. But Vanguard would be a purely scientific effort with minimum funding and support from the Defense Department. The Vanguard satellite would be launched using a small three-stage rocket to be developed especially for that purpose. It could not use any hardware developed for the military missile program.

In mid-February 1957, USAF General Bernard A. Schriever, who then headed the Air Force ballistic-missile program, was a featured speaker at the nation's first Astronautics Symposium, held in San Diego and jointly sponsored by the USAF and General Dynamics/Convair. During his talk, he said that rockets being developed for ICBMs would enable the U.S. to

move out into space. The next day Schriever received a telegram from the Pentagon, he later recalled, "saying that from
now on we were forbidden to use the word 'space' in any of
our speeches." Less than eight months later, the Russians
would unwittingly help American space enthusiasts lift that
embargo.

Even in mid-1957, only weeks before the Russians launched
Sputnik-1, talk of "space" was taboo except in the privacy of
the Pentagon and even there it got a cool reception. Schriever
later summed up the prevailing pre-Sputnik atmosphere in
these words: "I can recall pounding the halls of the Pentagon
in 1957, trying to get $10 million approved for our [USAF]
space program. We finally got the $10 million, but it was
spelled out that it would be just for component development.
No system whatsoever."* But after the first Russian satellite
was launched, Schriever spent most of his time traveling from
his West Coast headquarters to Washington to "testify before
Congress and talk to people in the Pentagon about why we
couldn't do things faster to get on with space," he later recalled.

One additional possible reason for the official taboo on
space talk was the growing CIA interest in surveillance satellites. By the mid-1950s the Lockheed U-2 aircraft was being
readied for its secret missions over the USSR. But those responsible for the bold project had to face the prospect that the
Russians eventually would discover the penetration of their
much-prized Iron Curtain and would then develop means to
knock down the aircraft. Sound long-range planning dictated
that work begin on a successor to the U-2, to cope with its
eventual demise. The reconnaissance satellite had much to
commend it as the successor to the U-2 if its many difficult
technical problems could be solved.

* *Air Force/Space Digest* magazine, May 1964, p. 161.

9

THE PROMISE, AND PROBLEMS, OF SPACEBORNE RECONNAISSANCE

An earth satellite offers several unique advantages for reconnaissance. One is its inherently high speed of nearly 18,000
mph, more than thirty-five times that of the U-2, which enables the spacecraft to survey extremely large areas in a brief
time. Another is its high vantage point. From an altitude of 200
miles, a satellite can photograph an area of several thousand
square miles, the size of Connecticut or even Massachusetts,
in a single frame of film.

Once in orbit in the near vacuum of space, a satellite can
"coast" for days, weeks or even years without propulsive fuel.
Increased altitude provides longer orbital life because of the
lower atmospheric density. However, for maximum photo
resolution of ground targets, satellite altitude should be as
low as possible. Also the lower the altitude, the larger the
possible payload that can be orbited with a rocket of any
given size. At the time of the initial RAND studies, it was
believed that an altitude of 300 miles would be required to
achieve an orbital life of a week or two. But by 1958, data
from early scientific satellites revealed that the density of the
atmosphere was less than expected and that an altitude of

approximately 100 miles would be sufficient for short-term missions, permitting much improved photo resolution.

The first photos taken from space revealed another unexpected advantage. The pictures proved to be much clearer than those taken from aircraft because of the absence of engine vibration and lack of atmospheric disturbance.

Still another important advantage of satellites for reconnaissance is that as the spacecraft circles the globe the rotation of the earth beneath brings much, or all, of the earth's surface into view. For a satellite whose orbit crosses the equator at an angle of 65 degrees, called the "inclination," all of the earth's surface between 65 degrees north latitude and 65 degrees south latitude will pass under the spacecraft's path. For any particular area, the satellite will pass over once each day while the region is in sunlight and again 12 hours later when the same area is in darkness. If the spacecraft is launched into a due-north/south orbit, then the entire earth's surface, from pole to pole, will pass beneath the satellite's path. This is termed a "polar orbit."

The time required for a satellite to circle the globe once, called its "period," depends on its altitude. At roughly 200 miles altitude, the period is approximately 90 minutes. If a satellite's altitude gives it precisely a 90-minute period, the earth will rotate to the east through an angle of 22½ degrees between each successive orbit so that to an earth-based observer each succeeding orbit would seem to shift 22½ degrees to the west. This "earth slippage" corresponds to a distance of 1,100 miles at a latitude of 45 degrees, and 1,565 mile at the equator. For example, if the satellite passed directly over Boston during one orbit, it would fly over Minneapolis on the next, and over Spokane, Wash., on the succeeding one. And because the 90-minute period is an even multiple of the earth's rotation period (1,440 minutes), the spacecraft would pass over these same three cities day after day.

For reconnaissance missions intended to locate secret military facilities, the spacecraft needs to gradually shift its pass-over point each day so that its cameras get a good nearly vertical look-down view at every square mile of earth below. This can be done by putting the satellite into an orbit whose period is *not* precisely an even multiple of the earth's rotation

period. For example, with a period of 90.5 minutes, a satellite's pass-over point will shift roughly 2 degrees to the west
each day, or 2 degrees to the east if the period is 89.5 minutes.
This corresponds to a shift of roughly 100 miles at a 45-degree
latitude. After approximately ten days in orbit, the spacecraft
would once again be passing directly over the same local
areas.

The sun's illumination angle over the region of interest,
i.e., the local time of day, can be very important for space
reconnaissance. For some missions, high noon may be desirable, while for others the long shadows of early morning or
late afternoon are helpful in subsequent photo analysis. The
sun angle is determined by the time of day that the satellite
is launched and its orbital inclination.

One of the most intriguing advantages of the reconnaissance satellite was that in the 1950s there was no existing
international law covering the question of a nation's sovereign
rights to the space high overhead. Was "outer space" to be
like the high seas, open to all users, or like the "airspace" immediately overhead, where each nation claimed sovereign
rights? The ambiguity of the situation was complicated by the
fact that with one notable exception (to be discussed in
Chapter 18), all conceivable satellite orbits necessarily required the spacecraft to pass over many national boundaries.

This issue remained a moot question until October 4, 1957,
when the Russians launched Sputnik-1 without permission or
notification of the many countries (including the U.S.) whose
national boundaries it over-flew. In a single unilateral stroke,
the Russians had unwittingly set in motion forces that would
shatter their Iron Curtain, thereby forfeiting their highly
prized secrecy.

While the reconnaissance satellite appeared to offer several important advantages, it simultaneously posed an even
more impressive array of very difficult technical problems
that first had to be solved. The most obvious, and most difficult, was how to get the photos back to earth. This question
had occupied the efforts of scientists both at RAND and in
industry for several years.

Two possible techniques were suggested. One was to equip

the satellite with a television camera and a tape recorder to store its TV pictures until the craft passed over a friendly ground station. At that time the satellite could transmit its stored pictures to earth by radio-link. In effect, the satellite would be a miniature television station with "delayed telecast." This approach was investigated in considerable detail by RCA during the early 1950s. The second approach considered was to use a more conventional camera with a special film that could be processed on board the spacecraft. Then, when the satellite came within range of a ground station, the processed film would be scanned by an "electric-eye" device to convert the photo into electrical signals which could be transmitted by radio-link. This approach was investigated by Eastman Kodak Co., assisted by CBS Laboratories, which developed the film scanner, and by Philco Corp. (now Philco-Ford), which devised the signal processor and radio link.

Both of these techniques suffered from one common disadvantage. If the picture achieved the desired ground resolution, each photo would contain a vast amount of detail from hundreds or thousands of square miles of earth surface. With the very limited amount of power then expected to be available on board a satellite, photo details could be transmitted at only a relatively slow rate. Theoretical analysis showed that if conventional aircraft-reconnaissance film were used, producing a 9 x 9 inch picture, it might take 20 minutes to transmit all the intrinsic detail on a single picture. Yet at low orbital altitude, a satellite would remain within range of a ground station for only roughly ten minutes per pass. (If a smaller 70 x 70 mm. frame size were used, transmission time could be cut to roughly two minutes, but with a corresponding decrease in ground resolution or coverage area.) Clearly, the reconnaissance satellite's great potential for rapidly photographing large areas in a short time faced a bottleneck if radio transmission were used, unless there were numerous suitably located ground stations.

By late 1954, studies conducted by RAND and by several industrial companies indicated that a reconnaissance satellite was feasible. On March 16, 1955, the USAF (under CIA sponsorship) issued a formal operational requirement for a Strategic Satellite System, identified by the code number

WS-117L. Three companies were selected for year-long design studies: The Martin Co. (now the Martin Marietta Corp.), Lockheed and RCA. After evaluating the results of these studies, the USAF notified Lockheed on June 30, 1956, that it had been selected to develop the satellite vehicle.

Lockheed's winning design concept called for building a second stage, to be mated with an Atlas ICBM first stage, which could propel itself into orbit and would carry a few hundred pounds of useful payload in its nose. The combination second stage and spacecraft would be approximately 19 feet long and 5 feet in diameter, with most of its volume used for rocket-engine fuel. The Agena, as the craft was subsequently named, would be powered by a 15,000-pound rocket engine developed by Bell Aerospace. The rocket originally was intended for use on a "flying bomb" to be carried by the B-58 supersonic bomber—a project which proved unworkable.

The Agena spacecraft could be used to carry any type of payload, but the reconnaissance mission was the primary one behind the initiation of the Lockheed effort. The very existence of the program was under heavy security wraps for more than a year, for two reasons. One was that the CIA was the sponsoring agency. The other was that some top administration officials, as noted earlier, were inclined to view satellites as "frivolous" or "fanciful." Officially the project carried the unclassified weapon-system designation of WS-117L, and was referred to as the Advanced Reconnaissance System. Lockheed itself used the code name of Project Pied Piper, whose derivation is unknown.

Typical of the many challenging problems that had to be solved was how to stabilize the satellite so that its camera would look vertically toward the earth. In a reconnaissance airplane, it is easy to determine the direction of "local vertical" and to cause the airplane to fly horizontally. A simple pendulum, which readily detects the direction of gravity, is combined with a gyroscope and connected to the aircraft's autopilot. But in a satellite in orbit, the force of gravity is canceled by the spacecraft's centrifugal force, resulting in the now well-known "zero-G" condition. In orbit a conventional pendulum is useless.

A similar problem arises in stabilizing a spacecraft about

its azimuth (heading) axis to assure that a series of photos taken during one orbit will be aligned parallel to another sequence taken during the next orbit, without any voids. For a reconnaissance airplane, the earth's magnetic field can be used to provide this azimuth stabilization function, but in orbit this handy reference is not suitable.

The ultimate solution would prove to be to use infrared sensors to detect the line of demarcation between the warm earth and cold space, and to split this subtended earth angle in half to determine the position of local vertical. These sensors are supplied by Barnes Engineering Corp. of Stamford, Conn. Azimuth stabilization would be obtained using another sensor to detect the position of the sun.

Another major problem was to achieve the necessary resolution of ground targets from orbit, at an altitude originally expected to be twenty times higher than anything previously attempted with an aerial camera. One of the immutable laws of photography is that resolution (i.e., the ability to detect and distinguish surface objects) is *reduced* as the distance between camera and subject increases, if other factors remain unchanged. The cameras carried by the high-flying U-2 had operated from altitudes of approximately 15 miles. If the same cameras could somehow be reduced sufficiently in size and weight to be carried in a satellite, without degrading their inherent resolution, there would automatically be a twentyfold reduction in ground resolution from an altitude of 300 miles. (Later it would be found that satisfactory orbital lifetimes could be obtained at an altitude of only 100 miles, thereby easing this problem.)

One way to compensate for the higher satellite altitude is to increase the focal length of the camera lens. But this increases size and weight, already a problem for the puny-size satellite payloads that then seemed possible. (Until December 1958 the largest satellite the U.S. was able to launch weighed only 38 pounds, of which less than 26 pounds was useful payload.) In the late 1940s the U.S. had developed the K-30 aircraft camera for oblique-angle photography from the side window of a high-flying airplane, to take pictures of airfields as much as 100 miles distant. But the K-30, with its 100-inch focal-length lens (using folded optics to reduce size), weighed

665 pounds and was bigger than a large office desk! Five more of the giant K-30 cameras were built in the early 1950s by Hycon Co., of Monrovia, Calif., and this subsequently led to the company's participation in the reconnaissance-satellite program.

Two other optical companies, which had been active in developing reconnaissance aircraft cameras, were also brought into the reconnaissance-satellite program by the CIA, which directs the development of the payloads. One of these was Itek Corp., of Lexington, Mass., an outgrowth of an Air Force-sponsored research group at Boston University, which developed an aircraft camera with a remarkable 240-inch lens. The other was Perkin-Elmer Co., of Norwalk, Conn., which had developed the first 48-inch airborne panoramic-scanning camera for the Air Force.

Beyond the dramatic reduction in camera size and weight for satellite use, the equipment had to be designed for *unattended* operation in the harsh, cold vacuum of space, for a period of days or weeks. The U.S. had gained a little experience in unattended aerial-camera operation in the late 1940s and early 1950s, when it had tried to use high-altitude balloons and prevailing winds to take photos behind the Iron Curtain. With luck, the balloon-borne cameras would eventually reach friendly shores, and the camera could then be recovered by parachute. But some also had come down before reaching friendly shores—which prompted Soviet protests.

In June 1956, the same month that Lockheed was selected to develop the reconnaissance-satellite vehicle, RAND scientists completed a report proposing a bold new idea for physically returning photos from orbit. The technique was especially attractive for obtaining very high resolution and/or color pictures.

The RAND report, classified "secret," bore the title *Physical Recovery of Satellite Payloads: A Preliminary Investigation.* It analyzed the problems of protecting heat-sensitive film from the searing temperatures that would be encountered in reentry into the earth's atmosphere at high speed. Techniques were then being developed for ICBMs to protect their nose-cones, containing a thermonuclear weapon, from the same ultra-high temperatures. RAND scientists proposed that a modified version of the ICBM nose-cone be carried by a reconnaissance satellite and used to return its precious cargo of film.

The technique was not as simple as it might sound. If the capsule were merely released from the satellite, it would remain in orbit. Instead, the capsule would have to carry a small rocket that could be fired to slow it down. As the capsule began to lose altitude, the more dense atmosphere would provide braking (and heating), causing a still further loss of altitude. Finally, upon reaching an altitude of roughly 50,000 feet, a parachute could be released to float the capsule down to earth—hopefully on the ocean, which would soften its impact. After landing, a small radio transmitter would automatically begin to broadcast a signal that could be used by search vessels to locate and recover the capsule. (Later, the idea of using specially equipped aircraft to snag the parachuting capsule in midair would be proposed and successfully developed.)

The recoverable film capsule was extremely attractive for certain types of missions but it posed many difficult technical problems. For example, unless the capsule and its retrorocket were correctly oriented at the time that the rocket was fired, the capsule would end up in a higher orbit instead of returning to earth. (This happened on more than one occasion, as will be described in a subsequent chapter.) Unless each event in a complex series occurred at precisely the correct instant, the valuable capsule would come down thousands of miles from waiting recovery forces. (This too would happen. In one instance, there is speculation that the capsule may have been recovered by the Russians.)

In August of 1957 there were efforts to accelerate the WS-117L program, but funding remained at a relatively low level. By this time, it had been decided to proceed with the radio-transmission-type camera system then under development by Eastman/CBS/Philco, pending the results of more intensive studies of the recoverable-capsule concept. The alternative approach of using a television camera had been dropped because it lacked the resolution needed.*

* The U.S. Army, which then had its own space ambitions, took over sponsorship of the TV-camera-type reconnaissance-satellite project at

On October 4, with the launch of Sputnik-1, "space" ceased to be a "dirty word," and administration officials who had previously all but ignored the WS-117L project suddenly found it an important focus of attention. The first public disclosure of the project appeared in the October 14, 1957, issue of *Aviation Week* magazine, under the headline:

USAF PUSHES PIED PIPER SPACE VEHICLE

Success of Soviet satellite will give new impetus to Lockheed project for reconnaissance satellite

The *Aviation Week* article reported that Lockheed had won the Pied Piper contract in 1956 and said that the reconnaissance-satellite program carried the unofficial nickname of Big Brother. (In George Orwell's famous book, *1984*, the closed-circuit TV cameras which enabled government officials to monitor the activities of all of the citizens, even in their homes, were fearfully referred to as Big Brother.)

According to the *Aviation Week* story, Eastman Kodak Co. and CBS Laboratories were believed to be major subcontractors. The satellite was expected to orbit at altitudes of 300 to 1,000 miles and to transmit its pictures to the ground by radio. There was speculation that the satellites might carry television, photographic cameras, infrared or radar sensors. At the then existing level of funding, approximately $12 million, the earliest possible date for launch of the first reconnaissance satellites appeared to be 1960. (By late November, Lockheed would receive a major increase in funding to accelerate the program.) The article said the program bore the official designation of WS-117L and the name Advanced Reconnais-

---

RCA for possible use in battlefield surveillance. The project was given the code name Tiros, an acronym derived from Television and Infra-Red Observation Satellite. By the fall of 1958 the Tiros satellite project would be reoriented to the more reasonable objective of transmitting TV pictures of cloud formations for meteorological use. Responsibility for the Tiros meteorological satellites was transferred to the newly formed civil space agency, NASA, and ultimately to the Weather Bureau. The meteorological satellite has played a key role in expanding scientific understanding of large-scale meteorological processes and in improving long-range weather forecasting.

sance System. (Later the Lockheed second-stage/spacecraft combination would be given the unclassified name "Agena," which is used regardless of the type of payload that is carried.)

In November 1957, RAND published another "secret" report, entitled *A Family of Recoverable Reconnaissance Satellites.* The report on a more detailed investigation of the feasibility of recovery from orbit was extremely optimistic over this approach to reconnaissance satellites. By January 1958 it had been decided to test the recoverable-capsule concept as soon as possible. This could be done most quickly by adapting the Agena so that it could be launched by the more available Thor IRBM, instead of waiting for the Atlas, even though less payload could be carried.

This is apparent from a censored transcript of testimony before the Senate Preparedness Investigating Subcommittee during the second week in January 1958. USAF Chief of Staff General Thomas D. White, speaking of the military uses of space, said: "Reconnaissance is . . . probably one of the earlier developments that will take place. You will have a vehicle that will map enormous areas of the earth frequently and perhaps by television and other means get the actual photographs down to earth."

General Schriever, who directed the USAF's missile and space effort, disclosed that "there was a lot of interest at different sources in the government for an advanced reconnaissance system"—a cryptic reference to the CIA. He also disclosed for the first time publicly that the Thor IRBM would be used to launch "a satellite whose photographic film payload would be physically recovered from orbit." Schriever continued: "Now since Sputniks, there has been, of course, a desire to accelerate this program and we have been looking at means for accelerating it." He said he had "given verbal instructions, and this will be carried out in contractual terms, to bring into this program the Thor as a booster to expedite getting orbiting vehicles and we think, based on our studies to date, and we have made rather exhaustive studies both in-house and in Lockheed, that we can get before the end of this year [1958] . . . perhaps as early as July, but more likely about October, we can get an orbiting vehicle with the Thor as a booster . . ."

When Senator Symington asked for an estimated date when a reconnaissance satellite might be available, Schriever replied: "I think that we can have a reconnaissance capability, using the Thor booster, by the spring of next year [1959], with a recoverable capsule . . ." Subsequent events would show that Schriever had underestimated the technical difficulties that lay ahead, but not grossly.

A very prescient observation was made during the same hearings by Lieutenant General Clarence Irvine, known to his intimates as Wild Bill, then Deputy Chief of Staff for Materiel. Senator John A. Carroll (Dem., Colo.) asked Irvine whether it was not likely that the Russians would also be able to launch a reconnaissance satellite, in view of their present strong capability in space technology, and the USAF official answered in the affirmative.

Senator Carroll then asked: "And by having such photographic cameras functioning, perhaps each of us would know what each other is doing; is that possible?" General Irvine replied: "I think this would be very healthy. This is the first step toward peace."

10

YEAR OF FRUSTRATION, MOMENT OF VICTORY

The Thor intermediate-range ballistic missile was still a largely unproven rocket in late 1957, when it was drafted for the additional mission of boosting experimental recoverable-type reconnaissance satellites into orbit. The first successful launch of a Thor had occurred only four months earlier and the first full-range flight had come on October 24. On November 25, 1957, the Pentagon had quadrupled the original $12-million funding level of the WS-117L program at Lockheed. By early 1958, Pentagon officials indicated that more than $150 million would be available for the program in the new fiscal year beginning July 1.

When the Thor was mated with the original version of the Lockheed Agena second stage, the vehicle could orbit 400 to 500 pounds of useful payload into a polar orbit. The Atlas/Agena could loft roughly six times as much payload, but every available Atlas was sorely needed to help close the Missile Gap. (The Thor would prove to have a far longer life as a space booster than originally expected as a result of more powerful rocket engines and the use of solid-propellant rockets strapped onto its main frame.)

The development of a complex, highly classified military

system is in some way like pregnancy. For many months after the program begins, there are few outward signs of progress. By late 1958 the most overt sign of progress in the military satellite program was a new set of nomenclature. The name "Pied Piper" had been dropped and the operational version of the reconnaissance satellite was now renamed Sentry. The experimental recoverable satellites to be launched by the Thor/Agena were given the family name of Discoverer.

Still another class of military-satellite function had appeared on the scene, called Midas—an acronym derived from Missile Detection And Surveillance. The Midas satellites were intended to detect a mass ICBM attack, shortly after the missiles were launched, which could provide nearly twice the fifteen-minute warning time available from the BMEWS radars in the Arctic. Midas was expected to be able to do this by detecting the infrared radiation emitted from the hot rocket plumes of the missiles. (The Midas satellite program, its many problems and eventual success, will be described in Chapter 18.)

Behind the scenes, several important new development efforts had been started. General Electric, one of two companies that had been working for several years to devise techniques to protect ICBM nose-cones from the multi-thousand-degree temperatures of hypersonic reentry, had been selected to apply these techniques to the recoverable capsule. (See Plate 4-B.) All-American Engineering Co., of Wilmington, Del., had been chosen to develop equipment which, it was hoped, would enable an Air Force airplane to snag and recover a capsule as it was parachuting to earth. The company would apply skills it had developed earlier in designing equipment for Navy aircraft carriers to enable airplanes landing on the deck to hook onto cables and come to a quick stop. Meanwhile, Eastman Kodak's reconnaissance camera, with its built-in film processor, and the associated CBS Laboratories' electronic scanner/converter were being readied for tests aboard an aircraft.

The major U.S. facility at Cape Canaveral, Fla., was poorly located for launching a satellite into polar orbit. If the rocket were fired in a due north or south direction, the booster might fall on inhabited areas. However, an ideal site for polar orbit launches was found at Point Arguello on the West Coast, approximately 150 miles northwest of Los Angeles, on terrain that jutted out into the Pacific and permitted a due-south launch without risk to inhabited areas. The facility was Cooke Air Force Base, subsequently renamed Vandenberg AFB.* During 1958, work began on the construction of launch and instrumentation for the Thor/Agena and subsequent launches of the Atlas/Agena.

On January 21, 1959, barely a year after the reconnaissance-satellite program had been given a high-priority green light, a Thor/Agena stood poised on its pad at Vandenberg. The Agena second stage measured 19.2 feet in length and 5 feet in diameter. Loaded with fuel, the Agena weighed more than 8,000 pounds, but once in orbit its "dry" (fuel-consumed) weight would be approximately 1,300 pounds, including a 245-pound reentry capsule.

Even before the countdown reached zero, a procedural error by a member of the launch team forced the mission to be aborted. A second launch attempt on February 25 also was aborted. But three days later, on February 28, the first Discoverer satellite was launched into successful orbit, with a perigee altitude of 99 miles, an apogee of 605 miles and an orbital plane inclination of 96 degrees. Once in orbit, something went awry with the spacecraft's attitude-stabilization system, and the satellite began to tumble crazily. This prevented any attempt to recover the capsule, but still it was a promising beginning.

On April 13, Discoverer-2 was placed in orbit with a perigee of 152 miles, an apogee of 225 miles and an inclination of 90 degrees. (The perigee altitude is the important parameter for reconnaissance because it sets the limit on satellite lifetime and target resolution.) When a radio command was sent to turn on the satellite's attitude-stabilization system, the craft responded obediently. It shifted to a tail-first orientation by using small jets of gas to generate thrust, then the satellite maintained its orientation toward the earth as intended.

The critical capsule ejection was scheduled to occur during

* More recently the facility has been given the new title of the Western Test Range, but the Vandenberg AFB terminology will be used here in future references.

the seventeenth orbit, but a human error caused the ejection signal to be transmitted prematurely. However, the retrorocket fired and the capsule's parachute apparently deployed properly, for it was reportedly seen descending near the northern tip of Norway, not far from its border with the Soviet Union. But a searching party was not able to find the capsule, despite its brightly colored parachute. This prompted speculation that the capsule had been recovered by a quickly organized Soviet search party. Despite the disappointment over failure to recover the capsule, the program appeared to be making remarkably good progress and there seemed good reason for optimism over the prospect of quickly achieving an operational capability.

By May 1959, the USAF had outfitted a few of its Fairchild C-119 cargo-transport planes with experimental versions of the All-American Engineering Co. equipment designed to snag and recover parachuting capsules in midair. These aircraft and their crews, part of the newly created 6593rd test squadron which was stationed at Hickham AFB, Hawaii, were busy practicing midair recovery using dummy capsules dropped from other aircraft flying at higher altitudes. The dummy capsule, like the one it simulated, carried a small radio beacon which began to transmit after the parachute deployed to enable the aircraft to "home" on the signal. By early June, the C-119 crews had become moderately proficient at the "airborne snatch," as it was called.

On June 3 the Discoverer program ran into its first bad luck when the satellite failed to go into orbit. Discoverer-4, launched on June 25, also failed to achieve orbit. Discoverer-5 broke the streak of launch failures on August 13, with a slightly heavier Agena spacecraft which weighed 1,700 pounds (dry) in orbit and included a 195-pound reentry capsule. All equipment performed well, including capsule ejection. But when the capsule retrorocket fired, a misorientation of the capsule caused it to be thrust into a higher orbit. This killed all chances of a controlled reentry and recovery.

Several weeks earlier, an ordinary Thor IRBM had been launched from Cape Canaveral, but it carried an unusual nose-cone on its 1,500-mile ballistic trajectory. This nose-cone, built by General Electric, was really a secret testbed for techniques to be used in the Discoverer program. The nose-cone, unlike a conventional IRBM design, carried a three-axis stabilization system, as well as a 16-mm. camera to photograph the earth so that GE scientists could tell how well the system performed. The nose-cone also was equipped with a Discoverer type of parachute and radio-beacon so that it could be recovered from the ocean at the end of its long trajectory, to enable scientists to study its photographs. The Thor nose-cone carried two infrared sensors, which would be used to determine the center of the earth (i.e. local vertical) by detecting the edges of the warm earth which contrasted with the very cold background of space. A third sensor would detect the position of the sun to orient the nose-cone in azimuth.

The official USAF announcement of the successful flight of this unusual nose-cone, following its recovery, made no reference to its relationship to the Discoverer program. The photos released, taken from an altitude of about 300 miles, had relatively poor definition. With the naked eye, one could see clouds and the rough outline of the Florida coast, but little more.

On August 19, 1959, little more than three weeks after the Thor flight from Florida, Discoverer-6 was put into orbit. In the past, the payloads had been identified as specific types of experiments. For Discoverer-6, the Air Force said the capsule contents were "classified," suggesting that the capsule contained a small camera, probably the first flown in the Discoverer series. Telemetry signals received from the spacecraft indicated that all was operating well. The capsule ejected according to plan, but no homing signals were received from its radio beacon. Without such guidance, there was no hope of finding and recovering the capsule and its "classified" contents.

During the next several months, GE and the USAF began a concerted program aimed at improving the reliability of each of the many components, every one of which had to operate properly to assure successful recovery of the capsule. Finally, on November 7, Discoverer-7 was launched into orbit. But an electronic failure on board the Agena itself prevented the craft from achieving all-attitude stabilization, causing the craft to tumble in space. This meant there was no hope of successfully ejecting and recovering the capsule.

On November 20 another attempt was made when Discoverer-8 was placed in orbit. However, a malfunction of the spacecraft guidance system resulted in an apogee altitude of 1,056 miles, much higher than planned. This required that capsule ejection be rescheduled from the seventeenth orbit to the fifteenth orbit, which would, hopefully, bring the capsule down only slightly south and east of the originally scheduled impact area. Despite these problems, the capsule was successfully ejected and its retrorocket fired properly. After firing, the rocket portion of the capsule is intended to separate to make way for the parachute to subsequently pop out. This separation occurred too late, according to radio telemetry signals received, causing the capsule to come down in still a different area from the one most recently planned. Telemetry signals also hinted that perhaps the parachute itself did not deploy. In any event, there was no capsule recovery.

To increase the chances of spotting and recovering the capsule, it was decided to add several new provisions. One was an intense rotating stroboscopic light, which could be seen by aircraft and surface-recovery crews if the capsule had to be recovered at night due to unexpected events, as with Discoverer-8. Another was to add "chaff"—thousands of tiny strands of metal foil—which could be released during descent and easily spotted by radar, to help locate the capsule's flight path.

By February 4, 1960, the capsule of Discoverer-9 had been outfitted with the additional recovery aids, and the satellite was ready for launch at Vandenberg. This time, a malfunction of propellant-loading equipment caused premature shutdown of the Thor first-stage rocket and no orbit was achieved. On February 19, Discoverer-10 was poised for launch. The Thor took off smartly, but at an altitude of 20,000 feet it veered off course and had to be destroyed by the range officer. Finally, on April 15, Discoverer-11 made it into a good orbit, with a perigee of 109 miles, an apogee of 380 miles and an inclination of 80 degrees. Its capsule contents were not disclosed. Presumably it contained a small camera pack. The capsule was ejected as planned but was not detected descending into the intended recovery area, despite all of the additional recovery aids on board.

It was now almost a year since the early successes of Discoverer-1 and -2 satellites had prompted great optimism over the feasibility of the recoverable-capsule concept. But the first capsule had yet to be recovered. Skepticism was growing in the Pentagon over the basic feasibility of the idea. There were, it seemed, simply too many events in the sequence, all of which had to function correctly and precisely on time, and too many opportunities for one or more things to go amiss.

In an effort to pinpoint weaknesses in the recovery-system design, the USAF turned to its Arnold Engineering Center at Tullahoma, Tenn., which had giant vacuum chambers that could simulate the cold rigors of space. There it was discovered that the capsule retrorocket's long exposure to the extremely low temperatures of space could adversely affect its performance. Also wind-tunnel tests suggested an improved parachute design. These and other design improvements were promptly introduced into Discoverer-12. On June 29 it was ready for launch. But this time a malfunction occurred in the attitude-control system of the Agena during launch, and no orbit was achieved!

With so long a string of failures, there were at least a few who suggested, only half in jest, that there ought not to be a Discoverer-13 launch attempt—the satellite should simply be redesignated Discoverer-14. But on August 10, 1960, ignoring all superstitions, Discoverer-13 was launched into a nearly perfect orbit. Its initial perigee altitude was 153 miles, its apogee was 375 miles and its inclination angle was 93 degrees. So far, so good.

In the darkened USAF Satellite Test Control Center near Lockheed's plant in Sunnyvale, Calif., south of San Francisco, company scientists and Air Force officers monitored the spacecraft performance on display consoles. (See Plate 8-B.) As Discoverer-13 came over the North Pole on the following day in its seventeenth orbit, everything seemed to be working perfectly. The time was about 3 P.M. in California. A recently constructed USAF facility at Kodiak, Alaska, was poised to transmit the command to eject the capsule if the Sunnyvale center gave the word. It came, and at 3:11 P.M. Pacific Standard Time, the Kodiak station sent its brief coded command to the satellite. The spacecraft promptly returned a tele-metered acknowledgment that it was responding.

As the spacecraft telemetry data was received, it was clear that Discoverer-13 had ejected its capsule, the retrorocket had fired and at 3:26 P.M. the satellite parachute had de-ployed. Discoverer-13's capsule appeared headed for the in-tended recovery area, approximately 300 miles northwest of Hawaii. Surface ships on station in the recovery area and C-119 aircraft flying over it were alerted.

A heavy layer of broken clouds at 10,000 feet made visual sighting difficult, but the capsule was spotted and tracked by radars in Hawaii and by large airborne radars aboard early-warning aircraft flying in the area. Aircraft crews caught a brief glimpse of the bright-orange parachute, but were unable to attempt a midair recovery. By 4:05 P.M. (PST), the capsule had been spotted in the water by low-flying search aircraft, thanks to its intense flashing light. Within several hours a heli-copter, with the aid of a Navy frogman, lifted the precious cargo from the ocean and put it safely aboard the USS *Haiti Victory*. The first man-made object had been recovered from space, after more than a year of thwarted attempts. (See Plate 4-A.)

To the layman, unaware of the strategic import of the reconnaissance-satellite program or the inherent difficulties of recovery from space, the newspaper headlines of the Dis-coverer-13 success must have seemed "much ado about noth-ing." After President Eisenhower viewed the recovered capsule in Washington, he cited it as evidence that the U.S. "leads the world in the activities in the space field that promise real benefits to mankind." The Air Force emphasized that the capsule had not contained any "sensor equipment," i.e. recon-naissance cameras.

A few days earlier a bulletin published by the USSR Academy of Sciences had lauded the "peaceful" intent of Soviet rockets and satellites while criticizing the U.S. space program as "provocative." The Soviet article criticized the Midas and Discoverer programs and spoke of America's use of space for "open espionage." Earlier, following the launch of the first Tiros weather satellite on April 1, 1960, the Soviets had charged it with being a reconnaissance satellite—as it originally was conceived to be. But in reality the resolution of the Tiros television camera was far too poor to be able to detect ground targets, although it was quite adequate for providing pictures of cloud formations.

Another good launch and orbit were achieved with Discoverer-14 on August 18 as the Air Force sought to show that the recovery of Discoverer-13 had not been merely a rare stroke of good luck. As Discoverer-14 came over the Pole on its seventeenth pass the following day, telemetry data from the spacecraft showed everything was in order. It was shortly after 7:30 in the morning (Hawaiian time) when the Kodiak station transmitted the critical command for the spacecraft to eject its capsule. The response was prompt and obedient. In the Hawaiian recovery area, six C-119 "Flying Boxcar" recovery aircraft from the 6593rd test squadron were flying within the 60 x 200-mile target area while three others were cruising slightly outside. There were several additional supporting aircraft of other types, as well as several surface ships equipped with helicopters for surface recovery.

At 7:53 A.M., the first strong signal from the parachuting capsule was received by the recovery team. Captain Harold E. Mitchell, commander of one of the C-119s which was then closest to the capsule's location, began his preparatory maneuvers as he waited for the capsule to drift down from its 55,000-foot altitude. At 8:06 A.M., the first visual sighting of the bright parachute occurred. By the time the capsule was approaching Mitchell's altitude, with a sink rate of about 1,500 feet per minute, it was dead ahead and he started his first pass. The aircraft was just above the top of the capsule parachute as it closed on it, the trapeze-like cables hanging out from the rear of the airplane fuselage to wrap around the chute. But the recovery cables missed by a matter of inches.

Mitchell dropped down to around 10,000 feet and turned to make a second pass. This time he missed by a matter of several feet. With luck, there would be time to make one more attempt before the capsule penetrated and was obscured by a layer of clouds at an altitude of 7,000 feet. Mitchell banked his airplane steeply, throttled back as much as he dared, and once more headed for the capsule parachute at an altitude of about 8,500 feet. This time the trapeze-like mechanism found its target, looped under it, and the parachute collapsed and folded over—exactly as intended. Mitchell had hooked a big fish. Now all that remained was to carefully winch it into the open door at the rear of the C-119. Technical Sergeant Louis Bannick reeled in the catch at 8:23 A.M., approximately thirty minutes after the capsule beacon signal had first been detected.*

The first midair recovery of a capsule returned from space was now history. Spaceborne reconnaissance had taken a giant step.

* Captain Mitchell received the Distinguished Flying Cross for his accomplishment, and the members of the crew were awarded Air Medals.

11

THE TELLTALE
SATELLITE PHOTOS

It is a very closely guarded secret as to when the Discoverer capsules began to carry small cameras and which was the first to return photos taken over the USSR. The Air Force statement on Discoverer-13 had emphasized that it carried "no sensor equipment," but no such disclaimer was made for Discoverer-14 following its recovery on August 19.

The photographic payload that could be carried in the capsule was limited to less than 100 pounds and about 3 cubic feet. In view of the long string of recovery-system malfunctions, it would be logical for the Discoverer-14 capsule to be stuffed with instrumentation and telemetry equipment designed to report back on the inner workings of the complex recovery system, to enable engineers to pinpoint the cause of previous problems. However, it had been three months since the Powers U-2 had been shot down. During this period, the U.S. had been deprived of photos that would help it appraise the much-feared Missile Gap. Thus, there would be considerable pressure to put Discoverer capsules to practical use as soon as possible.

Columnist Joseph Alsop, in an article published several years later (December 23, 1963), wrote: "We could only guess before [about Soviet missile strength] . . . But after August, 1960, we began to *know* that the ICBMs were not there." Alsop paid tribute to his close friend Richard M. Bissell, Jr., who had played a key role in pushing the U-2 and reconnaissance-satellite programs at the Central Intelligence Agency. The occasion for the tribute was Bissell's departure from the CIA. If Discoverer-14's capsule did not contain a small camera, probably Discoverer-15 did when it was launched on September 13. The satellite achieved a good orbit for picture-taking, with a 130-mile perigee. But a malfunction of the craft's attitude-stabilization system caused the capsule to fall several hundred miles outside the planned recovery area. The capsule later was sighted on the water but very rough seas prevented its recovery.

By early September the reconnaissance-satellite program once again had been given a new name, Samos, an acronym derived from Satellite And Missile Observation System. More important was the new sense of urgency and priority that had been triggered by the two successful capsule recoveries, by the successful aircraft tests of the Eastman Kodak radio-transmission type of photo-reconnaissance equipment, and possibly by photos returned from a Discoverer capsule. Where previously the Samos program had been directed by the USAF's West Coast Ballistic Missile Division, now program management was placed directly under the Secretary of the Air Force in Washington. Brigadier General Richard B. Curtin, who had been responsible for several USAF satellite programs on the West Coast, was transferred to the Secretary's office to direct Samos. This would help coordinate the program with the CIA, which would be the principal customer for photos produced by Samos.

The Eisenhower administration requested additional money to accelerate the reconnaissance-satellite program, and Congress promptly approved. Defense Secretary Gates indicated that the Pentagon planned to spend at least $50 million of the additional funds, beyond the $200 million originally authorized for Samos in the fiscal 1961 budget.

Aviation Week magazine, reporting these developments in its September 12, 1960, issue, said that Samos would be able to carry cameras with focal lengths as long as 40 inches. The next week, *Aviation Week* reported that the Air Force had hurriedly requested industry bids on several new types of recoverable capsules. These would be larger than Discoverer capsules to enable them to carry cameras especially designed for space reconnaissance. The USAF was seeking delivery of prototypes in nine months—half the normal development timetable for such complex hardware. By early October the Pentagon had decided to spend an additional $34 million on Samos, beyond the extra $50 million it had approved only a few weeks earlier.

The Discoverer capsule, for all its remarkable achievements, was not well suited to the nation's most pressing need in the fall of 1960: determining how many ICBMs were deployed behind the Iron Curtain in the vast reaches of the USSR. Any camera that could be squeezed into the Discoverer capsule could not carry a very big film supply, and ground resolution would leave much to be desired. This explains why, during the first week in October, attention focused on Vandenberg AFB, Calif., where a giant Atlas ICBM, mated to an Agena second stage, was being readied for launch. It would carry a radio-transmission-type photo-reconnaissance system developed by Eastman Kodak Co. With the much greater thrust of the Atlas, a useful payload of nearly 3,000 pounds could be orbited—roughly six times the weight of the original Thor/Agena. And the payload could be lofted into higher orbit, providing a useful life of many weeks instead of only a few days.

The added payload capacity would enable the spacecraft to carry a moderately large supply of film and a relatively long focal-length lens capable of providing good ground resolution despite the higher orbital altitude. The official Air Force announcement on the purpose of the mission was brief and cryptic. It was to "test photographic and related equipment . . . to determine engineering feasibility of obtaining an observation capability from an orbiting satellite." On October 11 the Atlas/Agena carrying Samos-1 took off on what seemed to be a successful launch, but the spacecraft failed to go into orbit. It would be more than three months before another Samos launch would be attempted.

Approximately two weeks later the Air Force introduced

an improved second stage, called Agena-B, for the launch of Discoverer-16. The Agena-B was 25 feet in length, roughly 6 feet longer than the original version, to accommodate more fuel and permit a longer thrusting period. This would roughly double orbital payload, raising it to nearly 1,000 pounds. The improved Bell Model 8048 rocket engine used new self-igniting (hypergolic) fuel, to simplify engine construction and improve reliability. The new Agena-B would be used with both the Thor and Atlas.

Following the first successful trial of the new Agena-B, the USAF launched a heavier Discoverer-17 into orbit on November 12. Its in-orbit ("dry") weight was 2,100 pounds, compared with 1,700 pounds for the earlier Agena-A. The satellite was allowed to remain in the cold near-vacuum of space for nearly two days before reentry was attempted on the thirty-first orbit. The satellite followed the radio commands from the Kodiak station, and the parachuting capsule was first sighted at an altitude of 30,000 feet as it approached the planned recovery area. Captain Gene W. Jones, flying one of the C-119s, recovered the capsule on his second attempt at an altitude of 9,500 feet. The official USAF announcement on the capsule payload said it contained "biological specimens." However, these would have occupied only a tiny fraction of the volume available in the capsule. It seems safe to assume that the capsule also carried camera equipment.

Meanwhile, the Air Force was outfitting five of its larger Lockheed C-130 transport aircraft for midair recovery, to handle the heavier second-generation reconnaissance capsules then under development which would be too big for the C-119s. The first of the new C-130 recovery aircraft would go into service early in 1961.

On December 7, Discoverer-18 was launched using an improved Thor whose thrust had been increased by 10 percent, to 165,000 pounds, to handle still-heavier payloads. A good orbit was achieved, and this time the spacecraft was allowed to remain in space for nearly three days before reentry was attempted. When the command signal was transmitted from the Kodiak station, the satellite responded like a well-trained automaton.

One of the waiting C-119s recovered the capsule at 14,000 feet, on the first attempt. Discoverer-18 was termed "the most precise and successful flight of a Discoverer vehicle to date," by Major General Osmond J. Ritland, commander of the USAF's Ballistic Missile Division. The spacecraft had used, for the first time, a new gyro stabilization system built by Honeywell. It was the second successful midair recovery in a row, and the third out of the last four in which the capsule had returned. After more than a year of frustration, the recoverable-capsule technique seemed to have been mastered.

Discoverer-19, launched on December 20, did not carry a recoverable capsule. Instead, it was equipped with instrumentation to measure infrared radiation from ballistic missiles and from the earth for use in the Midas satellite project. (See Chapter 18.)

Samos-2 was being readied for launch at Vandenberg in the hope that it would be the first radio-transmission-type reconnaissance satellite to survey the USSR and resolve the Missile Gap question. There was some uneasiness because of the earlier malfunction of the Atlas/Agena booster for Samos-1 and of another the previous February which was intended to launch Midas-1. The new booster had, however, launched Midas-2 into a near-perfect orbit in May, but on December 15 the Atlas (teamed with a different second stage) had failed to launch a lunar probe from Cape Canaveral.

The originally scheduled Samos-2 launch date slipped from December into January of 1961 as technicians checked and double-checked the booster and its complex satellite payload. Finally, on January 31, the countdown was under way. Shortly after noon the 99-foot-high Atlas/Agena took off. At 2:55 P.M. PST, the Air Force announced that Samos-2 was in orbit. It was a good one, with a perigee of 295 miles, an apogee of 343 miles and an inclination of 95 degrees. Samos-2 was revolving around the earth once every 95 minutes.

If the Eastman Kodak camera was equipped with a 40-inch telephoto lens and used film with a resolution capability of 100 lines per millimeter, from an altitude of 300 miles it could provide ground resolution of better than 20 feet. This would have been adequate to spot the giant SS-6 Soviet ICBMs and their launch-site support facilities. If the lens had a 10-degree field of view, it could photograph 2,500 square miles (50 x 50 miles) in a single picture. On this basis, the entire 8.6 million square miles of the USSR could be photographed with fewer than 4,000 pictures, allowing a slight overlap between each. But since much of the Soviet Union is uninhabited and inaccessible to railroads or modern highways needed to transport the huge Russian ICBMs, perhaps only one-quarter of the country would have to be surveyed to locate missile launch sites. On this basis, no more than 1,000 satellite photos would be needed to inventory Soviet ICBM strength.

Barely a month after Samos-2 went into orbit, its transmitters were turned off without official explanation. During this period, the satellite had made approximately 500 orbits, sufficient for it to have transmitted more than a thousand photos to several ground stations that had been drafted into service for this purpose. Several months would then be needed to analyze these hundreds of photos.

Four months later, in June 1961, the U.S. officially reduced its national intelligence estimate of the number of operational Soviet ICBMs by 50 percent! Instead of the 120 missiles that had been forecast to be operational by the summer of 1961, the figure was cut to only 60 missiles. (This was occurring as Khrushchev was confronting Kennedy with his Berlin ultimatum.) By September 1961, the official count of Soviet ICBMs would be slashed even more sharply, for there would be additional satellite photos.

Discoverer-20 had been launched into orbit on February 17, 1961—the heaviest yet, with an in-orbit weight of 2,450 pounds, thanks to the increased thrust now available from both the Thor and Agena-B. But spacecraft-stabilization problems arose to prevent any attempt to eject and recover the capsule. The next day, Discoverer-21 was placed in orbit. It did not carry a recoverable capsule because it was intended to make infrared measurements for the Midas early-warning satellite project. The spacecraft became the first to re-start its Agena-B engine briefly while in orbit, to increase satellite altitude. This would be used in later spacecraft to permit the lowest possible initial altitude, for good photo resolution, then to subsequently boost satellite altitude to avoid premature atmospheric entry.

Discoverer-22 was launched on March 30, with a recoverable capsule, but it failed to achieve orbit because of a malfunction. Another attempt was made on April 8 and Discoverer-23 went into a good orbit. But once again there were problems with the spacecraft-stabilization system. When the capsule was ejected, it ended up in a higher orbit instead of returning to earth. Discoverer-24, launched on June 8, failed to make orbit. Once again the Discoverer program was being plagued with troubles.

Finally, on June 16, 1961, Discoverer-25 was placed in a good orbit. Its capsule ejected on the thirty-third orbit, but not exactly according to plan. The capsule landed north of the expected recovery area but was finally recovered by Air Force frogmen, who parachuted into the ocean to save the capsule. This considerable effort to recover the Discoverer-25 capsule suggests it contained something more valuable than "biological specimens."

As the potential war clouds gathered over the Berlin crisis, the pace of Discoverer launches accelerated. Discoverer-26 was launched on July 7 and successful midair recovery of its capsule was achieved. But Discoverer-27, launched two weeks later, failed to achieve orbit when its Thor booster veered off course after take-off and had to be destroyed. Again, on August 3, Discoverer-28 did not get into orbit when the second-stage Agena malfunctioned.

Discoverer-29 achieved orbit on August 30. Although its capsule came down outside the planned recovery area, frogmen parachuted into the ocean and secured the capsule to a life raft until it could be picked up by the USS *Epperson*. On September 12, Discoverer-30 achieved orbit, and after two days in space its capsule was recovered in the air by a newly equipped C-130 airplane. Discoverer-31 was launched on September 17, but when the capsule-ejection command was sent, the spacecraft obstinately refused to comply. On September 9 an attempt to launch Samos-3 failed when its Atlas/Agena booster exploded on the pad.

By mid-September of 1961, the U.S. had recovered three Discoverer capsules in recent months, all of which presumably had carried small cameras, and it also had had time for a more detailed analysis of the Samos-2 pictures. As a result, the U.S. had once again revised its official count of the number of operational Russian ICBMs. *Now the figure was a mere 14 missiles.* This was less than one-quarter the previous figure established in June (as columnist Alsop correctly reported in his article of September 25, 1961). It contrasted sharply with the fearful estimate in December 1959 that 400 Russian ICBMs would be operational by mid-1961!

While the Russians had only 14 ICBMs that could strike at the U.S., America had three times as many Atlas ICBMs in operational readiness, plus 80 Polaris missiles, 600 SAC B-52 bombers and even more shorter-range B-47s.

Satellite photos taken of the Russian experimental-missile test facility at Tyuratam confirmed earlier intelligence obtained from U-2 pictures that revealed that the Soviets were developing a new generation of ICBMs—the SS-7 and SS-8. These smaller ICBMs, made possible by a reduction in weight of Soviet H-bombs, were better suited to hardened sites and fast reaction. The discovery of the new generation of Soviet ICBMs served to explain why the Russians had decided to build only a relative handful of the giant SS-6s, and encouraged U.S. officials to believe the small inventory of Soviet ICBMs that had been turned up by satellite pictures.

The Soviet SS-6 missiles were so large that they could be transported only by railroad and this meant that operational launch sites necessarily had to be situated near a major rail line—the Trans-Siberian railroad. When satellite photos revealed this constraint on Soviet missile deployment, it greatly reduced the total area that had to be reconnoitered within the USSR in the search for ICBMs.

In the early fall of 1961, as Khrushchev blustered and threatened over Berlin, confident that the U.S. government believed a Missile Gap in the Soviet favor existed, the Kennedy administration's problem was how to let the Russians know that the "cat was out of the bag." It would be a tragedy if the two giant powers bumbled into a thermonuclear war, even if the U.S. could easily "win" it—if such a word is appropriate.

On October 6 the President had returned from his vacation at Newport, R.I., to meet with Soviet Foreign Minister Gromyko. The two-hour discussion on the Berlin crisis offered a splendid opportunity for the President to tell the Soviet official that the U.S. now had satellite photos that revealed exactly how few ICBMs the Russians really had. The President also could suggest that these satellite photos enabled Strategic Air Command pilots and missileers to know precisely where these vital Russian targets were located, if the need should arise. To assure that Gromyko accepted these statements at face value, the President would have been tempted to show the Russian official several satellite photos of key Soviet air bases and ICBM installations. *Under the circumstances, it would have been extremely surprising if the President did not do so.*

Within eleven days of the Kennedy-Gromyko meeting, Khrushchev surprised the world and the 22nd Party Congress delegates by announcing that he was withdrawing the critical December 31 deadline for achieving a peace treaty with East Germany, effectively easing the Berlin crisis that had been boiling for months. Khrushchev's words, used to explain his action, are interesting in retrospect. He said that the Western powers now were "displaying a certain understanding of the situation . . ." If most of the delegates in Moscow failed to grasp the full significance of these words, at least a few top Soviet officials fully understood.

The coming months would require an agonizing reappraisal in the Kremlin, resulting in a daring attempt to restore a more favorable balance of power for the USSR which would once again bring the world to the brink of thermonuclear holocaust in October of the following year.

12

THE MISSILE GAP
TURNS ON THE USSR

By mid-fall of 1961, Soviet leaders were fully aware of the new American reconnaissance-satellite capability as a result of the President's own efforts and cryptic public statements by Gilpatric and Rusk. At that point, the President ordered the heaviest possible secrecy cloak on spaceborne reconnaissance, with a security classification of "top secret."

Russian military officials had known of the existence of the Samos and Midas programs at least since 1957–58, if only from reading open American publications. In a Russian book entitled *Science and Technology in Contemporary War*, by Major General G. I. Pokrovsky, which went to press on October 2, 1957 (two days before Sputnik-1), the author had discussed briefly the use of satellites for reconnaissance.*

One understandable reason for the new cloak of secrecy was to try to hold down discussion of the Samos program in the press to avoid an open affront to Russian sensibilities which might provoke them into developing means to destroy

---

* In discussing man-made satellites, Pokrovsky wrote: "These satellites, in addition to their scientific value, also have certain military significance; by means of them, one can conduct observation of the enemy's territory and drop atomic bombs on that territory."

U.S. spacecraft. Another reason was to withhold any details that might enable the Russians to devise passive counter-measures, such as camouflage. By mid-November 1961, the Samos and Midas programs had "never existed" insofar as open government publications were concerned.* The USAF was only allowed to announce that it had successfully launched an "unidentified" satellite. For those working on the programs, each type of satellite would be referred to only by code number, such as 188, 239.

By the spring of 1962 the U.S. and USSR would agree to report every satellite launch to the UN, giving its orbital characteristics but without the requirement to identify the satellite's intended function. Even with this meager data, plus a few additional details published by Britain's Royal Aircraft Establishment, which operates under less restrictive secrecy, it is often possible to deduce the function of an unidentified satellite.

For example, if a satellite goes into an orbit with a very low perigee and remains aloft for only a few days before being returned to earth (or ejecting a capsule), and if such spacecraft are launched at moderately regular intervals into such orbits, then the satellite probably has a "close-inspection"-type reconnaissance mission, and the film-pack is returned. If a satellite is launched into a slightly higher orbit, but with a perigee low enough for good ground resolution, and re-mains aloft for many weeks, then it probably is a radio-transmission type on a "search-and-find" mission. The first few Midas early-warning satellites went into near-circular orbits at roughly 2,000 miles altitude and were identified by name and function. Subsequent launches of "unidentified" satellites into 2,000-mile near-circular orbits—much too high for photo reconnaissance—leave little doubt as to their intended function.

A satellite may not achieve the intended orbit or respond to command signals once in orbit, so false inferences may occasionally be drawn about an individual spacecraft. But once the "bugs" are worked out, a moderately consistent pat-

---

* The Pentagon's telephone directory still listed a "Director of the Samos Project" until it was reprinted a few weeks later.

tern emerges to help identify the intended function, as we shall see.

The first attempted launch of an "unidentified" satellite, after secrecy was invoked by the U.S., came on November 22, 1961, using an Atlas/Agena booster which had been used previously for Samos. The satellite failed to orbit—which prevented positive identification of its intended function. However, it had been nearly a year since Samos-2 had made a full surveillance of the USSR and it would be time for another such mission. An attempt had been made to launch Samos-3 on September 9 (before the secrecy order), but the satellite failed to achieve orbit.

The U.S. government had decided against trying to retroactively convert the Discoverer program to "nonexistence" because it had been too widely publicized. Rather, it would officially "peter out" in the spring of 1962. Seven more of these were scheduled for launch, through Discoverer-38, which would be orbited on February 27, 1962. Of these seven attempts, five would achieve orbit and four of these would successfully eject their capsules. Of the four ejected capsules, three would be recovered in midair and the fourth from the ocean surface.

By the time the Discoverer program officially ended in early 1962, it had achieved a remarkably good record, considering the then relatively primitive state of the space art and the difficulty of the recovery task. Out of 38 attempted launches, 26 Discoverer satellites had gone into orbit. Of the 23 that carried separable capsules, 12 had been successfully ejected and recovered—eight in midair and four from the ocean. During the last half of the project, after initial problems had been pinpointed, the recovery batting average had gone up significantly.

On December 22, 1961, the U.S. successfully launched its first "unidentified" satellite into orbit, using the Atlas/Agena booster. The satellite perigee altitude was 145 miles, its apogee was 467 miles and the orbital inclination was 90 degrees. With the 145-mile perigee, and the use of the Agena engine re-start to boost satellite altitude later in the mission, the spacecraft remained in orbit for nearly eight months. Presumably this was a radio-transmission-type satellite designed to make the first fresh reconnaissance of the entire USSR in nearly a year. Its 145-mile perigee, approximately half of the 300-mile altitude of the original Samos-2, would provide a twofold improvement in ground resolution if the same camera system and lens were used. If these had been improved during the intervening months, perhaps resolutions as good as 5 to 10 feet were obtained.

As the first pictures from the new satellite were being radioed down and rushed to waiting photo-analysts, Defense Secretary Robert S. McNamara was preparing an important document. It was his first detailed report on the state of the nation's defenses since he had taken over the reins of the Pentagon and it would accompany the new fiscal 1963 Defense Department budget request to be submitted in January (1962) to the House Armed Services Committee. McNamara's annual report, which came to be known as his Posture Statement, would assume growing significance with each passing year. Increasingly, the Posture Statement contents would be aimed at Kremlin leaders as much as to Congress and the American public. The copy of the McNamara Posture Statement made public contained all but a few classified details found in the original submitted to Congress.

The Posture Statement, released during the third week in January 1962, spoke softly but confidently of U.S. strategic power. It said the U.S. had a "reasonably precise calculation" of the number and location of Russian missiles and air bases— euphemistically referred to in Pentagon jargon as the "Soviet target system." The Defense Secretary said: "There is no question but that, today, our strategic retaliatory forces are fully capable of destroying the Soviet target system, even after absorbing an initial nuclear surprise attack." (This ability to retaliate even after an enemy struck first would become known as a "second-strike" capability.)

McNamara cautioned that the U.S. should expect "the Soviet Union will eventually build a large intercontinental ballistic missile force." He added that "we must concentrate our efforts on the kind of strategic offensive forces which will be able to ride out an all-out attack by nuclear-armed ICBMs in sufficient strength to strike back decisively."

For this reason the administration would emphasize the

Polaris missile and the new Minuteman ICBM which could be protected in deep, hardened underground silos from which it could be fired barely more than a minute after the command was given. In the new budget the administration was seeking $2.2 billion to complete development of Minuteman and to order an additional 200 missiles. This would bring the total Minuteman arsenal to 800 missiles. McNamara recommended against increasing the number of Polaris submarines beyond the 41 then authorized, which could carry a total of 656 missiles. (The Navy had wanted to increase the number of submarines to 45.) The administration asked for $232 million to continue the development of the new Skybolt ballistic missile, designed to be carried under the wings of USAF B-52 bombers and launched in flight. (This program would be canceled in late 1962.)

In total, the Kennedy administration was asking $51.6 billion for defense. This was approximately 20 percent more than the budget that had been submitted a year earlier by the outgoing Eisenhower administration, but much of the increase was intended to strengthen the Pentagon's ability to fight limited (non-nuclear) wars.

McNamara studiously avoided the kind of boastful "rocket rattling" that Khrushchev and other Soviet leaders had employed during the years when the Russians appeared to enjoy the long end of the Missile Gap. But to Kremlin leaders, who now found themselves on the short end of the Missile Gap, events taking place in the U.S. must have seemed to have ominous overtones.

For example, in May of 1961, President Kennedy had nominated General Curtis E. LeMay, then USAF vice-chief of staff, to move up when the incumbent chief of staff retired. The administration had taken this action with mixed emotions because McNamara and LeMay had disagreed sharply on certain matters. But it seemed a fitting final reward to an officer who had served his country well during World War II and in the early postwar years when he had shaped the Strategic Air Command into a keenly honed force. So LeMay had become the head of the USAF and a member of the Joint Chiefs of Staff, the principal military advisers to the President.

But to Soviet leaders, LeMay's promotion meant that one of the most "hawkish" of the American military officers was now in an even better position to influence U.S. policy. As Air Force Chief of Staff, not only might he trigger America's fast-growing arsenal of thermonuclear-tipped missiles, but his words of advice would be near the President's ear, or so it must have seemed. The Russians knew that in the early post-World War II years some Air Force officers had argued for a preemptive strike against the USSR, and LeMay was believed to have shared those views at that time.

During the 1962 May Day celebration in Moscow, Defense Minister Malinovsky warned Communist-bloc nations of the danger of a preemptive nuclear attack by the U.S. In an order-of-the-day issued to the Soviet armed forces, Malinovsky accused the U.S. of "hatching plans for a surprise rocket nuclear attack."

The Soviet Union's Armed Forces newspaper, *Red Star,* in its May 11 edition, charged that President Kennedy "heartily supported the doctrine of preventive war involving a surprise rocket nuclear blow against the Soviet Union." The article cited a recently published interview with Kennedy, written by Stewart Alsop, in which the President had indicated that in some circumstances the U.S. might be forced to initiate the use of nuclear weapons. This had been intended as a warning to the Soviets against the use of force in Berlin, and the President subsequently stated that he had no intention of suggesting that the U.S. might launch a preemptive strike against the USSR. But Soviet leaders found no reassurance in the later statement. The author of the article in *Red Star* sought to reassure readers by noting that "The problem of destroying enemy rockets in flight has been successfully solved in the Soviet Union." But this could not conceal the growing Russian concern.

If the Berlin crisis had cooled, the basic issues still remained unresolved. On April 26, Secretary of State Rusk had publicly stated that withdrawal of Allied troops from West Berlin was not subject to negotiation in the continuing U.S.–Soviet discussions. On May 3 the Soviet Union had accused the U.S. of having adopted an inflexible and intransigent position on the Berlin question.

As seen through Russian eyes in early 1962, the Soviet Union faced a fearful, fast-growing threat of thermonuclear devastation. A remarkably candid, and accurate, appraisal of the dimensions of this threat was disclosed in a book published in the USSR in mid-1962 entitled *Military Strategy.** The book, edited by Marshal V. D. Sokolovsky, and containing a collection of articles written by fifteen different officers, was the first such comprehensive summary of Russian military thinking published openly in nearly forty years.

The chapter called "Military Strategy of Imperialist States," with its summary of the existing and projected future American weapons arsenal, must have been frightening reading to those within the USSR who knew how few ICBMs and long-range bombers the Soviets then had. (No comparison of Soviet and U.S. military strength was included.) While the Soviet book contains conflicting views on the implications of new weapons (probably reflecting the vested interests of the different authors), one chapter warned that *"Nuclear strikes by strategic missiles* will be of decisive, primary importance to the outcome of a modern war. . . . The aggressive imperialist bloc is preparing a war which will involve general destruction of cities, industrial regions and targets, and communication networks, and mass annihilation by nuclear blows on the civilian population throughout the entire territory of the socialist countries."

Any Russian reader might understandably conclude that the U.S. would easily be in a position to carry out its plans in the not too far distant future. The book noted that as of early 1962 the U.S. had operationally deployed 108 Atlas ICBMs and 54 Titan-1 missiles. (This was a modest exaggeration since delays in missile-site construction had caused slippage from the originally published U.S. timetables.) By the end of 1962, the Soviet book reported, the U.S. would have 132 operational Atlas missiles, and by the end of 1963 there would be an additional 54 Titan-2 missiles. All of the Titans and a few of the Atlas missiles, it noted, would be protected in underground silos.

---

* Translation of this book, with an excellent analysis and annotation supplied by RAND Corp. scientists, was published by Prentice-Hall in 1963, under the title *Soviet Military Strategy.*

The book pointed out that 800 of the newer Minuteman missiles, with "increased invulnerability," were expected to be operational by the end of 1966. This would raise the total U.S. land-based ICBM inventory to more than 1,000. (Although the book did not say so, *if* the U.S. struck first, this land-based ICBM force was enough to destroy Russian population centers many times over.)

Additionally, the book reported, the U.S. then had six Polaris submarines on station, each armed with 16 missiles, and that the Kennedy administration had accelerated the Polaris program immediately after taking office. When the 41 submarines then authorized were on station by the end of 1966, they would confront the Soviet Union with 656 more thermonuclear-tipped missiles. Further, the U.S. then had more than 600 long-range B-52 bombers, each capable of carrying a bigger thermonuclear payload than an ICBM, plus 1,100 shorter-range B-47's and 60 supersonic B-58 bombers. (This also was a slight overstatement of the actual numbers.)

The Soviet book also reported that the U.S. had launched a program to develop a mobile medium-range ballistic missile (MMRBM) for use by North Atlantic Treaty Organization (NATO) countries. The U.S. plans, it said, called for deployment of approximately 500 of these new nuclear-tipped missiles in West Europe by 1966. (This program, like the Skybolt, was later canceled, but at the time the book was written it was an added part of the missile threat which was expected to confront the USSR.)

The ominous implications of all this, as seen through Russian eyes, could well have been summed up in a paraphrased version of the editorial that had appeared in the January 4, 1958, edition of Britain's *New Statesman* at the time when the U.S. seemed to be on the short end of the Missile Gap. Thus paraphrased, it would read: ". . . irrespective of any efforts which Russia may now make, the U.S. preponderance in advanced weapons has reached such an absolute stage that Russia's national survival will depend, until 196? at least, on 'American benevolence.'"

The U.S. had conceived the Polaris submarine as a means of protecting its retaliatory capability despite a surprise Russian attack. But Soviet strategists would see things quite differently, for such submarines could stealthily approach the Russian shores and then suddenly launch several hundred missiles which could devastate the country's national vitals. To U.S. strategists, the Polaris had always been a "second-strike" weapon, but to the Soviets it had a very tempting first-strike potential. In the U.S. strategy, the Minuteman's hardened underground silo was intended to assure survival of at least part of the missile force after a Soviet first strike. But to Russian planners, it seemed that the U.S. could strike first with Polaris, knowing that its Minuteman could survive Russian retaliation and could then be used to administer the *coup de grace* to the USSR.

Since the fall of 1961 the Kennedy administration had known that the Missile Gap was strongly in the U.S.'s favor. Why would it then request additional billions of dollars in early 1962 to build 200 more Minuteman missiles unless it was planning (or at least considering) a preemptive strike? Or so it must have seemed to Soviet strategists.

On March 16, 1962, Khrushchev had sought to gain some psychological advantage by announcing that Russia had developed a new "global missile," which could attack by indirect routes and thus evade the costly American early-warning BMEWS radars in the Arctic. A few years earlier, when the U.S. relied solely on long-range bombers, this would have prompted serious concern, but in 1962 it created no "waves" in the Pentagon. By mid-July, Khrushchev had taken still another tack. Now he was boasting about the Soviet Union's new defense against enemy ballistic missiles. Russia's new ABM, Khrushchev proclaimed, could "hit a fly in the sky."

Another, more direct indication of Soviet concern over the results of America's spaceborne reconnaissance was the disclosure that the Russians were now building concrete walls around their ICBMs to protect them from thermonuclear-weapon blast effects. This was first reported publicly in the July 26, 1962, issue of *The New York Times* by Hanson W. Baldwin, another columnist with good Pentagon sources.

The Soviets were not attempting to protect their giant first-generation SS-6 ICBMs in underground silos but were using concrete blast shields, Baldwin indicated. He said the number of Soviet ICBM sites in operational status or under construction "number considerably less than 100. Most of them stand in closely clustered, above-ground, open launching sites, where one large nuclear blast would knock out eight or more sites. The 'coffin' type of semi-hardened construction is just beginning." The new information, Baldwin wrote, "presumably gathered by electronic and communications intelligence and by satellite cameras and other devices, is considered quite reliable by Washington, although some details are, of course, missing. . . . There is confidence in Washington, as a result, that the United States today has both a quantitative and qualitative, or technological lead. . . ."

The new evidence of Soviet efforts to protect their missiles may have come from the radio-transmission-type satellite that was launched on December 22, 1961, or a similar one placed in orbit on March 7, 1962. Or it may have come from one of the recoverable film-pack-type satellites launched during the first few months of 1962.

On March 16, 1962, the Soviets had launched a satellite with a new generic name, Cosmos, which it was announced would perform various measurements and experiments in space. Its low-perigee orbit of 135 miles was similar to that of the radio-transmission-type reconnaissance satellite which the U.S. had launched only nine days before. On April 6, Russia announced it had launched Cosmos-2, to make space-radiation measurements. The most intense bands of Van Allen radiation are found at altitudes above 2,000 miles, yet the Cosmos-2 high point (apogee) was only 969 miles and its perigee was 132 miles—in the "reconnaissance region."

Cosmos-3 was launched on April 24 to make additional measurements in space, the USSR announced, but it too had a low perigee altitude of 142 miles. Then, two days later, Russia launched Cosmos-4 into a slightly higher orbit with a perigee of 185 miles. Three days after launch, the Soviets announced that the entire satellite had been returned from space and recovered! By early 1962, recovery from orbit was not a new achievement for the USSR. On August 20, 1960, it had recovered two dogs after a day in orbit in Sputnik-5. (This had come nine days after the first U.S. recovery of a Discoverer capsule.) On March 9, 1961, the Russians had done it again with a canine passenger in Sputnik-9. This was to test the recovery system that would be used later to recover the first human cosmonaut, Yuri Gagarin, on April 12, 1961. Within four months, on August 7, 1961, the Russians also safely recovered cosmonaut Gherman Titov after one day in space.

The Russians also had demonstrated a capability for space-borne photography. In October of 1959 they had sent their Lunik-3 spacecraft looping around the moon and had photographed its backside. The 35-mm. film had been processed in the spacecraft, scanned electronically, and then transmitted over a lunar distance to the earth. The camera had two lenses, one with an 8-inch focal length, the other a more useful 20-inch focal length. The resolution of the Lunik-3 photos was relatively crude, but part of that may have been due to the very low power available to transmit the pictures over so long a distance. Clearly, the Russians had the makings of a Samos-type radio-transmission reconnaissance satellite in the fall of 1959. When the Soviets decided to enter the spaceborne reconnaissance field, probably about the same period, it seems likely that they would use a modified version of the Lunik-3 system, as well as a recoverable reconnaissance type.

Cosmos-5, launched on May 28, 1962, with a very low perigee altitude of 126 miles, remained in orbit for a year. It could well have been a radio-transmission-type reconnaissance satellite. Then, on July 28, the Russians launched another recoverable satellite, Cosmos-7, into a good reconnaissance altitude with a 130-mile perigee. Four days later the spacecraft returned from orbit. Cosmos-8, launched on August 18, went into an orbit with a 159-mile perigee and would remain aloft for roughly a year. Which of these satellites performed according to plan and which, if any, malfunctioned is known for sure only to the Russians.

By the summer of 1962, Khrushchev's top military advisers had convinced him that the Soviet Union dared not risk its survival on "American benevolence." Perhaps the decision was precipitated by a brief event that had occurred in the Pacific on May 6, when an American submarine had fired a fully armed Polaris missile which exploded its thermonuclear warhead at the end of its 1,400-mile trajectory. To U.S. officials, this was merely part of the nuclear tests that had been triggered by the Russian action in abrogating the earlier moratorium. It was the first test of a complete long-range ballistic missile, including its warhead, on which the U.S. planned to put most of its chips. But the Russians, who earlier had used their nuclear and rocket tests for psychological warfare as well as weapon evaluation, might naturally see a broader implication to this Polaris shot than the U.S. government ever intended. Meanwhile, the Navy had begun flight tests on a still longer-range Polaris missile, known as the A-3, whose 2,500-mile reach would enable it to strike still deeper into the Russian heartland. Also, the Minuteman test program was proceeding rapidly and the first of the programmed 800 missiles would soon be going into hardened silos, deep underground.

The Russian decision to install intermediate-range ballistic missiles in Cuba was bold, but hardly original. Several years before, when the U.S. faced the prospect of a fearful ICBM gap, it had decided to build the less complex, less costly Thor and Jupiter IRBMs and install them in Britain, Italy and Turkey, to provide the time needed to bring along the Atlas and Titan. Now the Russians hoped to checkmate the American ICBM arsenal by putting their IRBMs in Cuba.

By September, Soviet ships were covertly bringing Russian technicians and construction material to Cuba to build the launch pads for the missiles that would soon follow. (On September 30, during talks in Moscow between Khrushchev and the Secretary of the Department of the Interior, Stewart Udall, the Soviet leader used the occasion to extend an informal invitation to President Kennedy to visit the USSR in the near future. This gesture, presumably, was intended to show that all was now sweetness and light between the countries. Also, it might be convenient to have the President on an extended trip in the USSR if the Russian missiles were discovered, because this could delay any American response.)

The Russians launched another recoverable satellite, Cosmos-9, on September 27 and returned it from orbit four days later. If Cosmos-9 carried a camera, its photos would have been reassuring to Russian strategists, for there was no evidence that the U.S. had discovered what was afoot in Cuba and was massing for an invasion or air strike. It was not until October 14, in belated response to repeated rumors of offensive missiles being installed in Cuba, that a U-2 was dispatched to photograph western Cuba from an altitude of 78,000 feet. By the early morning of October 16, President Kennedy was examining the photographic evidence that showed Russian technicians working feverishly to install IRBMs which could hold America's Eastern Seaboard cities as hostage.*

As the Kennedy administration weighed military action in Cuba and U.S. forces converged on Florida to prepare for that contingency, the Russians launched another recoverable satellite, Cosmos-10, on October 17. When the Russian satellite returned from orbit on October 21, its pictures could have shown Florida air bases bulging with newly arrived aircraft and a large U.S. Navy task force already assembled in the Caribbean. On October 20 the Russians launched Cosmos-11, a nonrecoverable-type satellite, into a slightly higher, longer-lived orbit similar to that used by American spacecraft that transmit their pictures by radio-link. If Cosmos-11 carried similar equipment, the Russians would have been able to watch the day-by-day build-up of U.S. naval strength in the Caribbean and air strength in Florida.

The President entered the Cuban missile crisis with a very precise inventory of Soviet strategic missile and bomber strength, thanks to U.S. satellite photos. On September 29 the U.S. orbited a radio-transmission-type satellite which remained aloft until October 14, and another of the same type on October 9 which stayed up until November 16, by which time the crisis had passed. Certainly the photos from these satellites would have been scrutinized for any indication of increased Russian activity at their military sites and air bases.

What role, if any, Russian satellite pictures played in convincing Kremlin leaders that the U.S. was prepared to go the limit, prompting the humiliated USSR to withdraw its missiles

---

* So far as is known, the Soviet missile build-up in Cuba was not discovered from satellite photos. Every available satellite photo was being used to inventory ICBMs in the USSR. Once it was decided to perform photo reconnaissance of Cuba, it was much quicker and easier to use the U-2, flying out of nearby Florida bases. Whenever photo reconnaissance can be performed safely by aircraft, they are used in preference to satellites because their lower altitude provides higher resolution and the mission can be staged more quickly and at lower cost.

from Cuba, probably is known only to a few Russian leaders. If Soviet satellite pictures did play a part in the decision, it would have been the second time in barely a year that space-borne reconnaissance had demonstrated its stabilizing influence.

The automatons-in-orbit, adolescent as their performance was at that stage, had kept the two giant thermonuclear powers from bombing into World War III at least once, perhaps twice.

13

THE SOVIET REACTION

What the reactions of the Russians would be when they discovered that U.S. satellites were photographing the USSR's secret military facilities had been the subject of debate at high U.S. levels for several years, and especially after the U-2 incident. The Soviets prized secrecy almost for its own sake. They had repeatedly protested when the U.S. made feeble attempts to take photos from high-altitude balloons in the late 1940s and early 1950s. When the Russians discovered the U-2 over-flights, they had worked hard to devise weapons to knock down the aircraft.

If the Russians had developed an anti-ballistic missile, as they claimed in the early 1960s, it could be adapted to an anti-satellite mission. A satellite is a much more vulnerable target than an ICBM because the spacecraft remains in orbit for at least several days, making it easy to predict when and where it will appear on each subsequent orbit. That would make interception relatively simple unless the satellite carried an engine to permit maneuvers in space. This has been one of the potential uses for the Agena engine's in-space re-start capability, but unless most of a satellite's precious payload is devoted to fuel, it can do relatively little maneuvering except to change altitude. Thus, American reconnaissance satellites were almost "sitting ducks" if the Russians should make a determined effort to knock them down.

Another possible countermeasure loomed after the Soviets began to launch manned spacecraft in 1961. These could be used to rendezvous with U.S. military satellites to inspect and, if necessary, to destroy them. For this reason, U.S. officials monitored Soviet space technology for evidence of a rendezvous capability. (In the early 1960s the U.S. itself launched a program to develop an unmanned "inspector satellite" in the event the Russians attempted to put nuclear weapons into orbit, but the project later was canceled.)

On July 23, 1961, during the Berlin crisis, the Soviet Union's Armed Forces newspaper, *Red Star*, charged that the Tiros-3 weather satellite and the experimental Midas-3 early-warning satellite, launched thirteen days earlier, were acts of espionage and aggression. Comparing these two satellites with the U-2, the Russian newspaper declared: "A spy is a spy, no matter what height it flies." The newspaper charged that the Tiros and Midas were "for reconnaissance of Soviet rocket bases and other objects, and reporting of weather conditions over Soviet territory." This, it declared, "shows that the Pentagon has not given up its plan for spying on the Socialist camp." The writer had mistaken Midas for Samos, but this was unimportant. Within four months, the Kennedy administration would invoke secrecy for both Midas and Samos.

In the spring of 1962 the UN created a Committee on Peaceful Uses of Outer Space to develop international agreement on rules for space activities. Soon afterward, the Russians launched their first recoverable reconnaissance-type satellite, Cosmos-4. On June 7 the Russians presented a nine-point proposal to the first meeting in Geneva of the new UN Outer Space Committee. The Soviet proposal called for a ban on the use of satellites for reconnaissance. By July 28, when the Russians launched another recoverable satellite, Cosmos-7, the Geneva meeting had recessed because the U.S. was not willing to accept the Soviet-proposed ban on satellite reconnaissance.

At the next meeting of the Outer Space Committee in September, Soviet delegate Platon D. Morozov complained about

"U.S. spy satellites." A few days later, on September 27, Russia launched Cosmos-9, a recoverable type, followed by another (Cosmos-10) on October 17, during the Cuban missile crisis.

On December 3, 1962, Morozov spoke before the UN General Assembly's Political Committee and charged that "Such [satellite] observation is just as wrong as when intelligence data are obtained by other means, such as by photography from the air. . . ." The following day, the Soviet delegate argued that the use of satellites "for the collection of intelligence information in the territory of foreign states is incompatible with the objectives of mankind in the conquest of outer space." Yet several weeks later, on December 22, Russia launched Cosmos-12, another of its recoverable satellites. It was the first of the recoverable type to stay aloft for more than three or four days, and it remained in space for eight days. This indicates that by the end of 1962 Soviet satellites had become sufficiently reliable to venture longer-duration eight-day missions which would become the characteristic orbital lifetime.

On March 21, 1963, after a three-month winter lull, the Russians launched another recoverable satellite, Cosmos-13, which stayed aloft for eight days. At about the same time, private talks were initiated between the U.S. and USSR delegates in the hope of finding grounds for compromise. But on April 16 the Russians broke off these talks and announced that they would submit a new eleven-point proposal to the UN committee. The new proposal also called for a ban on reconnaissance satellites. Committee discussions ended several weeks later in a complete impasse.

An indication of the U.S. position advanced during the long, fruitless discussions can be gained from a speech delivered by Leonard C. Meeker of the U.S. State Department, who was a delegate to the UN Outer Space Committee. Speaking at McGill University's Institute of Air and Space Law in Montreal, in April 1963, Meeker said: "International law imposes no restrictions on observation from outside the limits of national jurisdiction. Observation from outer space, like observations from the high seas or from the air space above the high seas, is consistent with international law . . .

observations from space may in time provide support for arms
control . . . If in fact a nation is not preparing surprise attack, observations from space could help us know this and
thereby increase confidence in world security which might
otherwise be subject to added and unnecessary doubts."
Meeker added that "the progress of science, to which the
Soviet Union itself has made dramatic contributions, decrees
that we are all to live in an increasingly open world."

During the same month that Meeker spoke in Montreal,
the Russians launched two more recoverable satellites, Cos-
mos-15 and Cosmos-16—the first time they had managed to
orbit two reconnaissance satellites in a single month. During
May, the Russians launched Cosmos-18, which stayed aloft
for nine days.

The Soviets had been able to move so rapidly because
they had adapted the 10,000-pound Vostok-manned spacecraft and its recovery system for the photo mission. The large
size of the craft permitted a big camera to be carried, but
one of the Soviet SS-6 ICBMs was required to launch so
heavy a payload. Although the Russians sorely needed every
possible ICBM to help them close the new Missile Gap favor-
ing the U.S., it is apparent that they also were anxious to close
the reconnaissance-satellite gap.

Although the U.S. published data on its arsenal of strategic weapons, including where its air and missile bases were
situated, Soviet military planners would not want to depend
on information supplied by a potential enemy; they would
want their own confirming pictures. Additionally, the Russians
would be anxious to learn how much detail on Soviet weapons
could be obtained by U.S. spacecraft, and Russian satellite
photos would provide a useful clue. Also, the Soviets would
want to use photo satellites to determine the location of U.S.
targets with a precision required for targeting their missiles,
and possibly to evaluate the effectiveness of Soviet camouflage
techniques.

By the summer of 1963, the Russians had launched and
returned a total of nine recoverable satellites from orbit. How
many of these were successfully retrieved on the ground is
known only to the Soviets. But it would be surprising if there
were not a few malfunctions which brought down some of the

satellites far from the available recovery teams, or turned the spacecraft into a fireball when the parachute failed to open.

The first public indication of a possible change in Soviet attitude on reconnaissance satellites appeared in the July 15, 1963, issue of *The New York Times* in a column by roving correspondent C. L. Sulzberger. Just a week earlier, Sulzberger reported, Khrushchev had talked with Belgian Foreign Minister Paul Henri Spaak, and the Soviet leader had been in a jovial mood as the two enjoyed a picnic along the Dnieper River. During discussions of a possible treaty banning nuclear tests, Khrushchev argued that on-site inspection was not really needed to detect underground tests. Then he added: "Anyway, that function can now be assumed by satellites. Maybe I'll let you see my photographs."*

Less than two months after this reported discussion, on September 9, 1963, delegates to the UN Outer Space Committee noted a curious omission in the remarks of the new Soviet delegate, Dr. Nikolai T. Fedorenko. *He made absolutely no mention of "spy satellites."* It was the first Soviet speech on peaceful uses of space in which reconnaissance satellites had not been raised as an issue and soundly lambasted! Four days later, in another conciliatory move, the Soviets withdrew their earlier objection to allowing space projects to be undertaken by private companies, such as the newly created Communications Satellite (Comsat) Corp.

It was during the same week in September that a White House spokesman on space matters publicly hinted that some of the Cosmos satellites were probably engaged in photo reconnaissance of the U.S., rather than being scientific satellites as the Russians claimed. The occasion was a panel discussion on space at the Air Force Association annual conven-

---

* Less than a year later, on May 28, 1964, Khrushchev was interviewed in Moscow by former U.S. Senator William Benton. Khrushchev had urged that the U.S. stop its "provocative" photo-reconnaissance aircraft flights over Cuba, intended to assure that Russian missiles had not returned. Benton quoted Khrushchev as saying that U.S. reconnaissance-satellite photos were quite adequate to check on Cuban military facilities and much less provocative. Then, according to Benton, Khrushchev added: "If you wish, I can show you photos of your military bases taken from outer space. I will show them to President Johnson if he wishes." Finally, Khrushchev jested: "Why don't we exchange such photos?"

tion in Washington. When a member of the audience asked the panel whether some of the Russian satellites might be engaged in military missions, such as reconnaissance, Dr. Edward C. Welsh, executive secretary of the White House's National Aeronautics and Space Council and one of the panelists, volunteered to answer the question.

Welsh pointed out that all Russian satellites passed over the U.S. and added: "It would be stupid to conclude that they have not been making observations." Welsh noted that the Soviet need for satellite reconnaissance was less pressing than for the U.S., because of America's open society, but he said: "They have been practicing, I am quite confident."

Only a few days before Fedorenko had revealed the new Soviet attitude, Stanford University in Palo Alto, Calif., was host to a three-day conference called "Open Space and Peace Symposium." The meeting was sponsored by the Hoover Institution on War, Revolution and Peace and the nineteen speakers discussed space reconnaissance from a wide range of viewpoints, from the technical to the philosophical. One of the most prophetic speeches came from Dr. Stefan T. Possony, then the director of the Hoover Institution's international political studies program. "Space henceforth will remain a constant *sputnik*, or 'companion,' of mankind and will allow humanity to achieve a full quantum jump in the production of relevant information [for peace]," Dr. Possony said. "Few of us appreciate the magnitude of this advance."*

During the first week in November 1963, it was announced that private discussions between the Soviet and American

---

* Ambivalent feelings continued to persist within the USSR, at least for several years. For example, in a *Handbook of Astronautics*, published in 1966 by the USSR's Ministry of Defense, the authors briefly discuss the Samos reconnaissance satellites and then observe: "Military circles in the U.S., disregarding the regulations of international law, launch dozens of spy spacecraft each year. These vehicles are engaged in reconnaissance operations mainly over the USSR and other socialist countries." Two years later, sharply divergent views were expressed by one of Russia's top civilian space scientists when he spoke in Denver, in October 1968, to the local chapter of the American Institute of Aeronautics and Astronautics. Dr. Leonid Sedov said that space technology was making a major contribution to "the solution of the cardinal problem of our time—preservation of peace."

delegates had resolved all of the outstanding differences on the peaceful uses of space. The Soviets had dropped their objections to "American spy satellites." The Russians now were operating their own. More important, by late 1963, at least a few Kremlin leaders had finally recognized, along with some of their American counterparts, that in an ICBM/thermonuclear age, secrecy and national security are not necessarily synonymous.

14

SECOND-GENERATION U.S.
RECONNAISSANCE SATELLITES

By the time the Russians finally dropped their formal objections to spaceborne reconnaissance in late 1963, the U.S. had introduced a second generation of photo satellites—both radio-transmission and the recoverable film-pack types. Their photos revealed that the Soviets were starting to build Polaris-type submarines and underground silos to house the new Russian SS-7 and SS-8 ICBMs. (The SS-7 is a two-stage missile, similar to but slightly larger than the Atlas, which the Soviets would display for the first time at the 1964 November Day parade in Moscow. The SS-8, a storable liquid-fueled missile similar to the Titan-2, would be unveiled at the 1965 May Day parade in Moscow.)

The introduction of a new generation of radio-transmission-type satellite in mid-1963 is indicated by the debut of a much more powerful booster to launch this type of spacecraft. Originally, it will be recalled, the Atlas/Agena had been used to orbit the Samos radio-transmission-type spacecraft, but by early 1962 Eastman Kodak had reduced the size/weight of the camera system sufficiently to permit it to be launched by using the smaller, less costly Thor/Agena. (The Atlas/Agena had been reassigned to carry heavier cameras with long focal-length lenses used with recoverable film-packs.)

The new launcher was the Thrust-Augmented-Thor (TAT), consisting of the basic Thor IRBM plus a cluster of three solid-propellant rockets (each generating roughly 50,000-pound thrust) attached to the main rocket body. When combined with the latest model Agena-D, with its longer-thrusting, slightly more powerful engine, the TAT/Agena-D could loft slightly more than 2,000 pounds of useful payload into a reconnaissance-type orbit, or nearly twice the payload of an ordinary Thor/Agena. This would permit the satellite to carry more film and batteries for a longer useful life in orbit.

The Agena-D, first introduced in the summer of 1962, employed a new engine design which enabled the rocket to be re-started in space—an accomplishment that is more difficult than it sounds. This re-start capability could be used, for instance, to apply thrust during the mission to raise satellite altitude and to extend its orbital lifetime.

The first attempted TAT/Agena-D launch on February 28, 1963, ended in failure, as did a second on March 18. The first successful launch to orbit a radio-transmission-type satellite came on May 18. During the rest of the year, the new booster lofted six more reconnaissance satellites of this type into orbit for search-and-find-type missions.

By 1964 the U.S. was launching radio-transmission-type satellites at roughly one-month intervals. Each would usually remain aloft three to four weeks before natural decay and destruction occurred, but sometimes orbital life was shorter, suggesting the possibility of a satellite malfunction which prevented the Agena engine from boosting vehicle altitude. Occasionally a second satellite would be launched while one already was in orbit, suggesting that the first may have malfunctioned or run dry of film. Sometimes there would be a brief gap in continuity, as if a planned launch had been delayed because of booster or payload problems.

During 1964, the U.S. successfully orbited a total of 12 radio-transmission-type reconnaissance satellites, all launched by a TAT/Agena-D booster. This type normally is launched in the early afternoon, between 1 P.M. and 3 P.M. local standard time. This means that areas to be photographed will be illuminated by the sun from an angle just below the zenith, and tall objects will cast a slight shadow.

U.S. RADIO-TRANSMISSION-TYPE RECON SATELLITES
ORBITED IN 1964

Launched	Decayed	Perigee	Apogee
Feb. 15	Mar. 9	119 mi.	278 mi.
Mar. 24	(Failed to achieve orbit)		
Apr. 27	May 26	109	277
June 4	June 18	93	267
June 19	July 16	109	287
July 10	Aug. 6	112	286
Aug. 5	Aug. 31	112	262
Sept. 14	Oct. 6	119	286
Oct. 5	Oct. 26	109	243
Oct. 17	Nov. 4	117	258
Nov. 2	Nov. 28	112	278
Nov. 18	Dec. 6	112	211
Dec. 19	Jan. 14	114	237

During the 1964 Presidential campaign, Republican candidate Barry Goldwater sharply criticized the Johnson administration for its defense policies, especially its restraint in weapons procurement. On September 22, Defense Secretary McNamara responded to this criticism by revealing the official, if approximate, count of Soviet ICBM strength. Speaking to the American Legion in Dallas, McNamara said the U.S. had "more than 800 fully armed, dependable ICBMs deployed on launchers, almost all in hardened and dispersed silos. *The Soviet Union has fewer than one-quarter this number, and fewer still in hardened silos* [emphasis added]."

An indication of U.S. confidence in the accuracy of its inventory of Soviet missile strength obtained from satellite reconnaissance came in January 1965. McNamara's Posture Statement disclosed that the administration had turned down the USAF's request to buy 200 more Minuteman missiles, beyond the 1,000 already approved. (The number would remain frozen at 1,000 after 1965.) The Defense Secretary noted that by the summer of 1965 the U.S. would have 800 Minuteman missiles deployed, plus another 464 Polaris missiles on station.

During 1965 the U.S. launched 13 of its radio-transmission-type satellites using the TAT/Agena-D, compared with the 12 orbited during the previous year, as shown below:

U.S. RADIO-TRANSMISSION-TYPE RECON SATELLITES
ORBITED IN 1965

Launched	Decayed	Perigee	Apogee
Jan. 15	Feb. 9	112 mi.	261 mi.
Feb. 25	Mar. 18	110	234
Mar. 25	Apr. 4	116	165
Apr. 29	May 26	114	291
May 18	June 15	123	206
June 9	June 22	109	221
July 19	Aug. 18	114	277
Aug. 17	Oct. 11	112	253
Sept. 22	Oct. 11	119	226
Oct. 5	Oct. 29	126	201
Oct. 28	Nov. 17	107	268
Dec. 9	Dec. 26	112	262
Dec. 24	Jan. 20	112	269

Although the U.S. has never made public even one photo taken by its reconnaissance satellites, it is possible to accurately estimate what sort of ground resolution was being obtained from radio-transmission-type spacecraft in the early 1960s. *The reason is that similar equipment was later used by the National Aeronautics and Space Administration for taking pictures of the lunar surface from satellites in orbit around the moon, and these pictures have been made public.*

The companies selected to furnish the camera-system components for the NASA Lunar Orbiters are the same ones that were making the functionally similar equipment for the strategic reconnaissance satellites. These include Eastman Kodak, which builds the camera and film processor; CBS Laboratories, which manufactures the film scanner/converter and Philco-Ford, which supplies the electronic-signal processing equipment.

Following the 1961 decision to try to land men on the moon before the end of the decade, NASA sorely needed photos of the lunar surface to help select landing sites which were both topographically suitable and scientifically interesting. This could be done by launching small photo-reconnaissance satellites into lunar orbit, processing the pictures on board the craft and then transmitting them back to earth by radio-link. There was scant time to develop and "debug" the required camera system for this mission and no need to try if the CIA would release its classified equipment for this purpose.

Agreement was finally reached, with the proviso that NASA would not release too many details on the camera, film processor and scanner/converter elements. The equipment authorized for NASA's use almost certainly was the first-generation design rather than the improved model which was introduced for strategic reconnaissance in 1963. Because the lunar satellites could orbit at very low altitude because of the absence of any atmosphere around the moon, NASA could use a shorter focal-length telephoto lens and still obtain adequate resolution for its purposes. The camera system carried in the 850-pound Lunar Orbiter spacecraft weighed only 145 pounds, including enough film for 400 photos. It was equipped with two lenses: one with a 3-inch focal length, and the other a 24-inch telephoto. Five of the satellites were successfully launched into lunar orbit between mid-1966 and mid-1967. The complex camera system performed perfectly—mute testimony to its previous years of service over the USSR and Red China. (See Plates 6-A, 6-B.)

The CBS Laboratories scanner/converter operated with a light beam measuring only two ten-thousandths of an inch in diameter (one-twentieth the diameter of a human hair). As the beam scanned back and forth across each frame of film, light, dark and gray areas on the picture were converted to corresponding electrical signals for transmission back to earth. Because of the limited electrical power on board the Lunar Orbiter, and the more than 200,000-mile distance over which the signals had to be sent, the pictures were scanned slowly and approximately 40 minutes were required to send back the contents of each frame of the 70-mm.-wide film.

The Lunar Orbiter camera system contained another feature used in strategic reconnaissance satellites—a means for automatically compensating for the high-speed motion of the spacecraft while the photo is being taken. Without such compensation, objects on the surface would have a smeared image, greatly reducing resolution. Compensation is provided by moving the film past the camera aperture at a speed which corresponds to the satellite's angular velocity over the surface, as measured by what is called a "velocity/height sensor."

Despite the modest supply of film that could be carried aboard each Lunar Orbiter, *the five satellites were able to photograph essentially the entire surface of the moon (including the backside), an area nearly twice as large as the entire USSR.* From an altitude of 28 miles above the lunar surface, each picture made with the 24-inch telephoto lens (which had a relatively small field of view) covered an area of approximately 43 square miles. But at an altitude of 112 miles, corresponding to that used for strategic earth reconnaissance, a single photo made by this camera would cover an area of nearly 700 square miles. And the lower-resolution (3-inch) lens, which could produce pictures capable of spotting large-scale construction efforts needed to build Soviet ICBM launch facilities, could capture more than 7,000 square miles of terrestrial surface in a single photo from an altitude of 112 miles. On this basis, all of the Soviet Union, including its vast uninhabited regions, could be surveyed with fewer than 1,200 photos made by the Lunar Orbiter camera system.

In terms of resolution the 24-inch telephoto lens, operating from a lunar altitude of 28 miles, produced pictures in which *objects smaller than 3 feet could be seen.* (See Plate 7.) Extrapolating by a factor of 4:1 for earth reconnaissance from a 112-mile altitude, the Lunar Orbiter camera could produce pictures with a ground resolution better than 12 feet. (This ignores possible degradation due to scintillation effects caused by the earth's atmosphere.)

If the early strategic satellites of the radio-transmission-type used a longer focal-length lens, such as the 40-inch telephoto lens, as reported by *Aviation Week,* then objects as small as 7 feet in diameter should have been discernible. This would have been adequate not only to examine details of Soviet ICBM launch facilities but also to tell whether a giant ICBM was in place at each site.

One of the major advantages of the radio-transmission-type satellite is its ability to promptly return reconnaissance photos, usually within one hour after being taken—providing suitably equipped ground stations are within range after the satellite passes over the area of interest in the USSR or Red China.

The USAF's network of ground stations for receiving satellite pictures and transmitting commands to the spacecraft, originally located along the West Coast of the U.S., has been expanded around the globe and now totals seven stations. These are located in New Boston (near Manchester), N.H., at Vandenberg AFB, Calif., on the Hawaiian island of Oahu, on Kodiak Island in Alaska, on Guam and on the British Seychelles Islands in the Indian Ocean, approximately 600 miles northeast of Madagascar. The seventh station is in an east African country which will not be identified to avoid possible embarrassment. Each station has at least one giant 60-foot-diameter antenna to receive picture signals.

Additionally, there are six shipboard stations, each outfitted with a 30-foot antenna which can be deployed around the globe as needed. Some of them also are used during missile tests to track the weapon. The first of these shipboard stations, the *General H. H. Arnold,* became fully operational in the fall of 1964.

For a typical radio-transmission-satellite orbit, with an inclination of approximately 80 degrees, the spacecraft's first *daylight* pass over Communist territory occurs at the eastern tip of Siberia. Soon the satellite comes within range* of the Guam station and can transmit down its photos. Two orbits later, the satellite passes over the east coast of Red China and shortly afterward comes within range of the New Boston, N.H., station. Pictures taken over central Siberia or Red China's missile test range may be radioed to New Boston, or perhaps a shipboard station in the Indian Ocean. As the satellite begins to pass over western Russia, where many of the most important military installations are located, many more photos will be taken and radioed down to stations at Van-

* Maximum range at which a station can receive transmissions from satellites depends on spacecraft altitude and the topography of ground-station location. For radio-transmission satellites, maximum range is typically about 750 miles.

denberg, Seychelles and in east Africa. Pictures taken as the satellite passes over Russia's Communist neighbors to the west will be radioed to the stations at Kodiak and Hawaii.

The received radio signals are recorded on magnetic tape at the station but are not reconstituted into pictures until they have been flown by special USAF jet courier aircraft to Washington and turned over to the National Photographic Interpretation Center. This facility, housed in the old Naval Gun Factory along the Anacostia River, is a 10-minute drive from the White House and the Pentagon and only a little more distant from the CIA's headquarters near McLean, Va.

For shipboard stations, the tape-recorded photos can be transferred to a specially outfitted aircraft by playing back and transmitting the signals to the airplane as it circles the ship. The signals received by the aircraft are recorded on tape and it is flown to a major air base for transfer to the courier airplane. Another technique which can be used during good weather is physical recovery of the tape by means of an adaptation of the midair recovery procedure employed for satellite capsules. The tapes are suspended in a capsule supported by a tall superstructure on the bow of the ship, and an aircraft flies low, drops its hook and snatches the capsule. (See Plate 8-A.) (A new and more speedy means for relaying satellite photos to Washington via communications satellite will be discussed in Chapter 17.)

A new generation of *recoverable*-type reconnaissance satellite, whose development had been initiated in late 1960, also made its debut in 1963. The portion of the satellite which housed the cameras and the recoverable film capsule was built by General Electric's Missile and Space Division in Philadelphia, which had originally devised the recoverable capsules for the Discoverer spacecraft. It is believed that the nonrecoverable part of the GE package contained several cameras which employed strips of film that fed out through a chute onto take-up reels in the 36-inch-diameter recoverable capsule, as shown in Plate 9-A. Unlike the Discoverer capsule, which had contained its own retrorocket, in the new design the re-start capability of the main Agena-D engine served to decelerate the spacecraft and initiate the reentry sequence.

By 1964 the new-generation recoverable reconnaissance satellites were being launched into orbit with almost clocklike regularity, lofted by the powerful Atlas/Agena-D.* This type of satellite usually remained in orbit for only three to five days before using up its film and returning the capsule to earth. Its function was to make high-resolution photos of missile sites and other strategic military and manufacturing facilities spotted on radio-transmission satellite pictures. Because only a brief orbital lifetime was needed and maximum possible resolution was wanted, this class of satellite was placed in orbit with a somewhat lower perigee than the radio-transmission type.

The following tabulations of recoverable satellites orbited during 1964 and 1965 show fewer launches during the winter months. One possible explanation is that much of the USSR experiences a harsh winter, which discourages major construction projects out of doors. Each of the satellites shown was orbited by an Atlas/Agena-D.

U.S. RECOVERABLE-TYPE RECON SATELLITES
ORBITED IN 1964

Launched	Orbit-Life	Perigee	Apogee
Feb. 25	5 days	107 mi.	118 mi.
Mar. 11	5	89	240
Apr. 23	5	93	209
May 19	3	88	236
July 6	2	75	215
Aug. 14	9	93	191
Sept. 23	5	90	188
Oct. 8	(Failed to achieve orbit)		
Oct. 23	5	86	168
Dec. 4	1	94	222

The recoverable satellites orbited in 1964 were launched at a slightly earlier time of day than the radio-transmission

---

* Satellite capsules also were being recovered with clocklike precision. By late 1963 the 6593rd Test Squadron was achieving midair recovery of 88 percent of the capsules, according to a United Press International dispatch. Howard Simons and Chalmers Roberts in an article that appeared in the December 8, 1963, issue of *The Washington Post,* quote an unnamed official as saying he "couldn't remember when they missed one."

type, presumably because more direct (zenith) illumination was needed for their close-inspection-type photographic mission. Launch time ranged from approximately 11:30 A.M. to 2 P.M. But by 1965 the launches were being timed much more precisely and were occurring a few minutes before noon so that the sun would be at its zenith relative to targets below the orbit. During 1965 the U.S. orbited a total of eight recoverable satellites, one less than in 1964.

U.S. RECOVERABLE-TYPE RECON SATELLITES
ORBITED IN 1965

Launched	Orbit-Life	Perigee	Apogee
Jan. 23	5 days	91 mi.	181 mi.
Mar. 12	5	93	178
Apr. 28	5	95	171
May 27	5	93	166
June 25	5	94	176
Aug. 3	4	93	191
Sept. 30	5	98	164
Nov. 8	3	91	172

There is no direct benchmark comparable to the Lunar Orbiter photos that can be used to estimate the resolution of photos obtained from the second generation of recoverable reconnaissance satellites. However, it is known that a giant camera, originally developed by Hycon for aircraft use, was tested in August 1963 aboard a very-high-altitude balloon, housed in an enclosure that resembled the nose section of an Agena spacecraft. The balloon was launched from an Air Force site in New Mexico. This would be a relatively inexpensive way to test camera performance before committing it to a costly spacecraft trial.

This LG-77 camera had a lens with a 66-inch focal length and used film with a 4.5-inch-square frame size. If film with an inherent resolution of 200 lines per millimeter were used, the Hycon camera should have been able to show objects as small as 2 feet in size from an altitude of 100 miles. The basic camera weighed nearly 400 pounds and could easily have been accommodated in the more than 4,000-pound payload capability of the improved Atlas/Agena-D.

In early 1965 the USAF began to launch another type of

"unidentified" satellite from its West Coast facility. These satellites went into a near-polar orbit similar to those of photo-reconnaissance spacecraft, but their perigee altitude of roughly 300 miles was too high for good resolution of ground objects. The first satellites were launched using a Thor combined with a small solid-propellant second stage called Altair. In the fall of 1966 a more powerful second stage, Burner-2, was introduced, and the mystery satellites were launched into a more circular orbit at an altitude of approximately 500 miles. Now their orbital characteristics began to closely resemble those of the civil Tiros meteorological satellite launched by NASA.

The first Tiros satellite had gone into orbit in 1960, the same year that the U.S. had started its Discoverer photo-reconnaissance experiments. The Tiros carried a TV-type camera to make low-resolution cloud-formation pictures, which were stored temporarily on magnetic tape until the satellite passed over a ground station and could transmit down the photos. For the first time, thanks to Tiros, weather forecasters in the Environmental Science Services Administration (ESSA) could obtain a synoptic view of the earth's cloud cover on a global basis—an invaluable aid in long-range weather forecasting.* Tiros pictures taken over the USSR and Red China also played a key role in the reconnaissance-satellite effort by enabling USAF spacecraft programmers to predict when specific areas of strategic interest would be free of cloud cover, to avoid wasting precious spacecraft film.

If the new mystery satellites which the USAF began to launch in 1965 were weather satellites, as their orbital characteristics suggested, why had it become necessary to deploy special military spacecraft for a function which Tiros had previously provided? The explanation is that reconnaissance-satellite programmers would need sufficient resolution to predict whether a very small area within the USSR or Red China would be free of clouds several hours later, whereas ESSA's civil meteorologists prefer large-scale cloud-cover pictures. Starting in early 1966, all future civil Tiros satellites would be

* In the fall of 1970, the Environmental Science Services Administration (ESSA) was merged with several other federal agencies to become the National Oceanic and Atmospheric Administration (NOAA).

launched into an 800-mile-high orbit, instead of the previous 500 miles, which would result in even less resolution.

This explains why the USAF had to start to provide its own meteorological satellite coverage in the mid-1960s. Presumably the USAF weather satellites are built by RCA and are similar to those which the company constructs for ESSA. But the USAF spacecraft must remain cloaked in secrecy lest some Congressman ask why the Air Force needs its own weather satellites, seemingly duplicating a function the civil Tiros could provide. To explain the need would involve disclosure of the reconnaissance-satellite effort, and this would "compromise" its "top-secret" classification.

15

MORE VERSATILE SENSORS
FOR SPACEBORNE RECONNAISSANCE

Even as the U.S. reconnaissance-satellite program was achieving a routine-like operational status in 1964–65, efforts were under way to devise more sophisticated sensors than the ordinary camera to cope with possible Soviet countermeasures. The Russians had prized their secrecy for decades, and old habits die hard. There would be great temptation for the Soviets to try to devise means to hide their strategic facilities from the prying eyes in space, probably by resorting to camouflage techniques first devised during World War II, some of which initially had proved quite effective.

Large portions of the northern regions of the USSR are often covered by clouds. Perhaps the Russians might try to construct underground missile silos during overcast periods, or at night using subdued light. During clear weather and daylight, such construction could be covered by camouflage. Once the silo was completed, the missile itself could be covertly moved into place under cover of clouds or darkness. The ICBM silo could be covered with a Hollywood movie-set type of structure, made to resemble some inconspicuous facility such as a bicycle-repair shop. This dummy structure would be designed so that it could easily be shifted if it was necessary to launch the hidden ICBM. Meanwhile, American photo-analysts could easily gloss over the structure if they even bothered to photograph it.

Or the Russians might attempt to construct dummy missile silos to draw the fire of American ICBMs in the event of war. In the eighteenth century, Russian Prime Minister Grigori Potemkin resorted to such trickery to fool Catherine the Great when she took a boat trip down the Volga to inspect her newly conquered lands. To create the illusion of a prosperous land, Potemkin built a movie-set type of village and had it propped up along the riverbank. Each night, after Catherine had passed the "Potemkin village," it was taken down and hastily moved to the next site along her route. Soviet Major General N. Talensky has publicly made a brief reference to the use of "sham targets" to draw American missile fire.

During World War II a technique was developed to penetrate camouflage which used photographic film that is sensitive to infrared radiation, which itself is invisible to the human eye. The principle on which it operates can be understood by considering the visible part of the spectrum. During daylight, live grass appears green to the eye, and to ordinary color film, because it absorbs all the colors in sunlight *except green,* which it reflects back. Dead grass appears brown because it reflects this color and absorbs all others. Sunlight also contains infrared radiation, called "near-infrared" because its wavelength is close to that of visible light. Some objects absorb this near-infrared almost completely, while others are much more reflective. Thus the amount of near-infrared reflected by any object is a characteristic "fingerprint" similar to its color in the visible region.

A piece of canvas that has been painted green to resemble grass might fool an aerial observer, or appear quite natural on ordinary color film. But if the picture is taken on film sensitive to near-infrared, the painted canvas will look quite different from ordinary grass because its reflectance of infrared will be markedly different. Since infrared-sensitive film has been used in aerial photography for many years, it certainly would have been used early in the reconnaissance-satellite program.

But during the early 1960s an even more exciting technique emerged—called "multi-spectral photography." Instead of taking only two photos of a scene, one with ordinary color film and one with infrared-sensitive film, a cluster of cameras could be used, each with a different filter so that it would photograph only the light reflected in a narrow portion of the visible spectrum. For example, one photo might be taken with a camera using a blue filter that would admit only blue light while a second would photograph the same scene with a red filter, and a third with a green filter. A fourth picture might be made with infrared-sensitive film.

After processing, the films can be viewed individually or in combination with one or more of the others, and they can be projected using color filters to enhance one particular spectral band. To the human eye, newly planted grass may look the same as long-established grass in an ordinary color photo. But with multi-spectral photography, new grass may be easily spotted because it reflects more light of one color while older grass reflects more of another. This would be especially useful in spotting newly planted grass intended to camouflage a missile silo. (See Plates 12-A, B, C.)

The first airborne tests of a multi-spectral camera system began in late 1962 aboard an Air Force C-130 transport. The equipment consisted of nine cameras, each filtered to photograph a different spectral band. The camera had been built by Itek Corp., under Pentagon sponsorship, in the hope it could detect clandestine underground nuclear explosions. At that time, all previous efforts to reach an agreement with the USSR to ban nuclear tests had faltered over the question of whether on-site inspection was needed to police the ban if it included underground tests. U.S. scientists speculated that such underground explosions might produce subtle changes in the reflective characteristics of the soil or foliage. The Pentagon had contracted with Itek to develop the nine-band camera to test the idea.

Whether the tests aboard the C-130, and those that followed in space during the 1964–65 period, demonstrated the feasibility of detecting underground nuclear tests has not been disclosed. But the tests clearly demonstrated the great reconnaissance potential of the technique. By late 1965, news of the exciting potentialities of multi-spectral sensors had spread

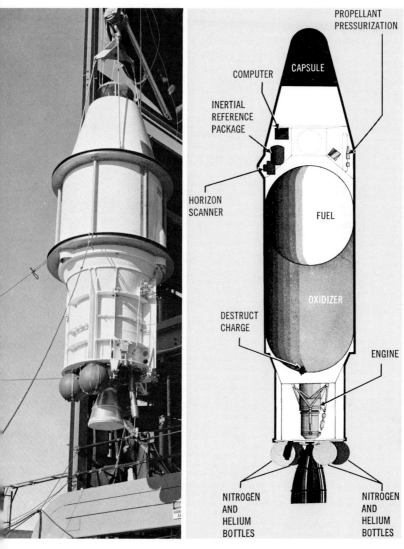

PROPELLANT
PRESSURIZATION

CAPSULE

COMPUTER

INERTIAL
REFERENCE
PACKAGE

HORIZON
SCANNER

FUEL

OXIDIZER

DESTRUCT
CHARGE

ENGINE

NITROGEN
AND
HELIUM
BOTTLES

NITROGEN
AND
HELIUM
BOTTLES

USAF's Agena spacecraft, built by Lockheed, has been used to orbit U.S.   1
photo reconnaissance satellites since the program began in the early
1960s. (See Chapter 9.) The Agena functions both as a second-stage
rocket, to achieve orbital velocity, and as a carrier for cameras and other
sensors located in the conical-shaped nose. Photo shows early Agena-B
being hoisted into position for firing tests. Longer, more powerful Agena-D
was introduced in mid-1962. Sketch shows early-model Agena's internal
construction, including equipment used to stabilize spacecraft attitude
for spaceborne photography. Model shown is one outfitted with recoverable nose capsule. Camera equipment is located directly behind capsule,
but not shown in this sketch.

Launch of a radio-transmission-type reconnaissance satellite aboard a Thor/Agena-D is shown in this series of photos. The McDonnell Douglas Thor intermediate range ballistic missile, originally drafted for temporary service as a first-stage booster, has since had its payload capacity greatly increased by expanding the Thor's fuel capacity and adding strap-on solid-propellant rockets. The current model, called the Long-Tank-Thrust-Augmented-Thor (LTTAT), is being used to orbit very heavy radio-transmission-type reconnaissance payloads.

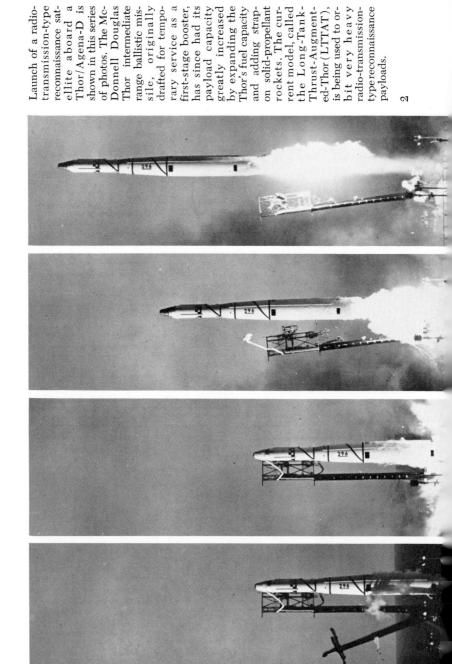

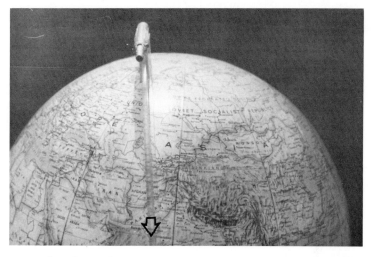

Near-polar orbit used for most U.S. reconnaissance satellites is shown by plastic girdle and small spacecraft model. For short-duration missions, satellite orbit remains essentially fixed in space while earth rotates so that most of earth's surface (except for the Poles) passes underneath the orbital path twice each day, once in daylight and once in darkness. Orbit shown has an inclination of 80 degrees.

With each successive orbit, spacecraft appears to move westward relative to an earth-based observer. For typical reconnaissance satellite missions, with an orbital period of approximately 90 minutes, the earth rotates through angle of approximately 22½ degrees between successive orbits as shown by "earth slippage" between Plates 3-A and 3-B. At the equator, this corresponds to a distance of roughly 1,565 miles.

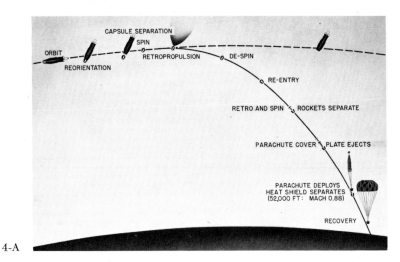

4-A

Return of reconnaissance films from space, a technique pioneered in the USAF's Discoverer satellite series, involves a complex series of events (shown in sketch above), each of which must occur at precise instant if capsule is to be successfully recovered in intended area. More than a year of failures preceded the first successful ocean recovery of Discoverer-13 on August 11, 1960, and the first midair snatch of Discoverer-14 on August 19, 1960. (See Chapter 10.) Plate 4-B shows construction of Discoverer capsule, developed by General Electric, which became prototype of larger reconnaissance film-pack recovery capsules introduced in early 1960s.

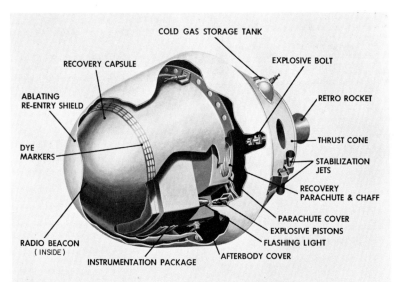

COLD GAS STORAGE TANK
RECOVERY CAPSULE
EXPLOSIVE BOLT
ABLATING RE-ENTRY SHIELD
RETRO ROCKET
THRUST CONE
STABILIZATION JETS
DYE MARKERS
RECOVERY PARACHUTE & CHAFF
PARACHUTE COVER
EXPLOSIVE PISTONS
FLASHING LIGHT
RADIO BEACON (INSIDE)
INSTRUMENTATION PACKAGE
AFTERBODY COVER

RE-ENTRY-RECOVERY VEHICLE

4-B

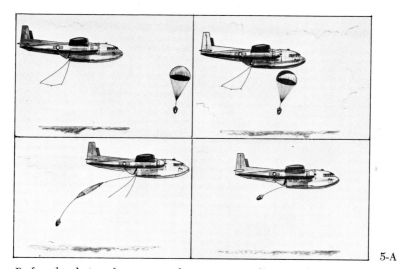

5-A

Preferred technique for recovery of reconnaissance film capsule is a midair snatch, using special trapeze-like equipment built by All-American Engineering Co. It was first tested on a Fairchild Hiller C-119 transport during Discoverer series. Currently, the larger Lockheed C-130 aircraft are being used to recover the bigger, heavier reconnaissance satellite capsules. Plate 5-B shows a C-119 attempting a practice catch of a dummy capsule dropped from a high-flying jet.

5-B

6-A

NASA

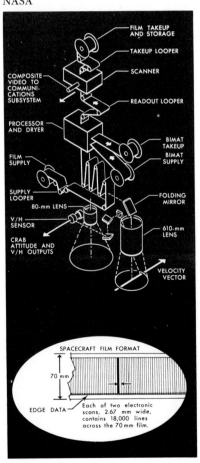

FILM TAKEUP AND STORAGE

TAKEUP LOOPER

COMPOSITE VIDEO TO COMMUNICATIONS SUBSYSTEM

SCANNER

READOUT LOOPER

PROCESSOR AND DRYER

FILM SUPPLY

BIMAT TAKEUP

BIMAT SUPPLY

SUPPLY LOOPER

FOLDING MIRROR

80-mm LENS

V/H SENSOR

610-mm LENS

CRAB ATTITUDE AND V/H OUTPUTS

VELOCITY VECTOR

SPACECRAFT FILM FORMAT

70 mm

EDGE DATA

Each of two electronic scans, 2.67 mm wide, contains 18,000 lines across the 70 mm film.

6-B

This 145-pound camera system, used in NASA's Lunar Orbiter satellites to photograph the lunar surface, is similar to equipment used in early Samos radio-transmission-type reconnaissance satellites. Film is processed by camera system (arrow), and each photo is then scanned by an electric eye to convert picture to electrical signals for radio transmission. (See Plate 6-B for details.) Camera system was built by Eastman Kodak Co., and the film scanner by CBS Laboratories, both of which supplied similar equipment for Samos satellites.

Lunar-surface photo, taken from an altitude of 100 miles by Lunar Orbiter-5, illustrates ground resolution obtained from earliest Samos reconnaissance satellites, which used similar equipment. Smaller of the two rocks (arrows), approximately 15 feet in diameter, is readily seen with the naked eye. Even narrower trail left by the smaller rock as it rolled down the hill illustrates ease of spotting roads used for construction of Soviet ICBM sites.

7

8A

One of six floating tracking-telemetry stations used by the USAF to receive pictures transmitted from radio-reconnaissance-type satellites which supplement ground-based stations in Guam, Hawaii, Alaska, California and New Hampshire. Pictures are received by giant 60-foot antenna near stern (arrow), while other antennas are used for radar tracking. (See Chapter 14.) Received pictures can be relayed to Washington for analysis by several techniques, including use of military communications satellites. Tape recordings of the received pictures also can be physically recovered during good weather by low-flying airplane, which snatches container from superstructure located near the bow (arrow).

8-B

Air Force Satellite Control Center, located near Lockheed plant in Sunnyvale, Calif., is responsible for monitoring and controlling reconnaissance satellites after they are in orbit. It is operated by the 6594th Test Wing of the USAF's Space and Missile Systems Organization (SAMSO).

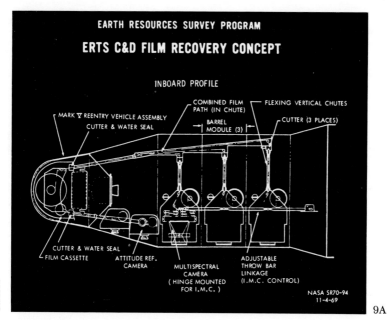

EARTH RESOURCES SURVEY PROGRAM

ERTS C&D FILM RECOVERY CONCEPT

INBOARD PROFILE

COMBINED FILM PATH (IN CHUTE) — FLEXING VERTICAL CHUTES

MARK Ⅴ REENTRY VEHICLE ASSEMBLY
CUTTER & WATER SEAL

BARREL MODULE (3)

CUTTER (3 PLACES)

CUTTER & WATER SEAL
FILM CASSETTE

ATTITUDE REF. CAMERA

MULTISPECTRAL CAMERA (HINGE MOUNTED FOR I.M.C.)

ADJUSTABLE THROW BAR LINKAGE (I.M.C. CONTROL)

NASA SR70-94
11-4-69

NASA

9A

Technique used to feed film from several cameras to recoverable capsule in second-generation reconnaissance satellites may have been similar to one shown in sketch. General Electric, which produced the second-generation recoverable capsules, submitted the sketch to NASA for possible use in the Space Agency's earth-resources survey program, where film recovery may be employed for certain missions.

Lockheed

9-B

Agena-D spacecraft, with much larger payload introduced in 1966, is shown undergoing final test at Lockheed plant in Sunnyvale, Calif., prior to shipment to Vandenberg AF Base for launch. (See Chapter 17.)

NASA

10   Excellent resolution of photos taken from space, even using ordinary short-lens cameras owned by more affluent amateur photographers, stems from lack of vibration and atmospheric turbulence experienced in pictures taken from aircraft. This photo, taken by Gemini-7 astronauts from an altitude of more than 100 miles with a Hasselblad camera, shows the Kennedy Space Center complex at Cape Canaveral, Fla. Rocket launch pads are visible to the naked eye (arrows), despite the obscuration from clouds. Original photo is in color, providing clearer resolution.

Photo of Dallas airport, taken by Gemini-5 astronauts from more than 100 miles' altitude, clearly shows runways, buildings and surface vehicles, despite "fuzzing" of the picture by the Defense Department to reduce definition. Picture was taken with a small telephoto lens, having a 48-inch focal length with folded optics, built for use by amateur astronomers rather than for aerial photography. (See Chapter 15.)

12-A

Multi-spectral camera, which takes several simultaneous photos of the same scene with each lens filtered to pass light from a different part of the visible and near-infrared spectrum, is especially useful in penetrating camouflage. Shown in first multi-spectral camera, with nine lenses, built by Itek Corp. and flight-tested on an aircraft in 1962. (See Chapter 15.) First operational use of multi-spectral cameras in satellites probably occurred around 1966.

Effectiveness of multi-spectral camera photos in spotting camouflage is shown by these two pictures taken from an aircraft over an Army camouflage test facility, where a number of military vehicles have been covered by camouflage nets. Plate 12-B is a black/white print of an ordinary color picture, while Plate 12-C is one of several multi-spectral pictures. Note that presence of camouflage, which is not readily discernible in Plate 12-B, stands out boldly in Plate 12-C (arrows).

12-B

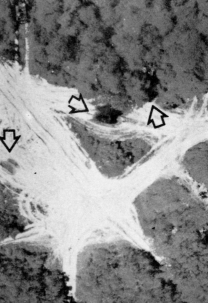

12-C

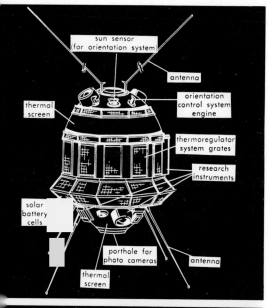

sun sensor
(for orientation system)

antenna

thermal
screen

orientation
control system
engine

thermoregulator
system grates

research
instruments

solar
battery
cells

porthole for
photo cameras

antenna

thermal
screen

Soviet Lunik-3, launched October 4, 1959, to take photos of the backside of the moon, processed the pictures on board and transmitted them back to earth, employing a technique similar to that used in America's radio-transmission-type reconnaissance satellites. The USSR is believed to have adapted the Lunik-3 spacecraft to military reconnaissance, both to monitor U.S. Navy deployment and to return cloud-cover pictures so that Russian recoverable reconnaissance satellites could be programmed to take pictures when areas of interest were free of clouds. (See Chapter 16.)

At least some of Russia's Molniya communications satellites are believed to contain a television camera, or other sensors, to warn of any U.S. missile attack. The Russian Molniya satellites, unlike American communications satellites which "hover" at a fixed longitude along the equator, are launched into a highly elliptical orbit that provides better coverage of northern portions of the USSR. In the highly elliptical orbit, the Soviet satellite remains relatively motionless with respect to the earth below for nearly eight hours out of every twelve. Once a day this apogee occurs over the USSR, where spacecraft can be used for communications. Twelve hours later, this apogee occurs over North America where satellite could perform early-warning mission. (See Chapter 18.)

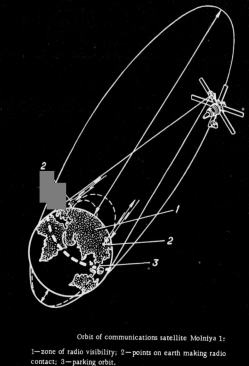

2

1

2

3

Orbit of communications satellite Molniya 1:

1—zone of radio visibility; 2—points on earth making radio contact; 3—parking orbit.

14

Texas Instruments Inc.

Infrared scanner can make a camera-like picture at night from thermal radiation emitted by all objects, making it useful for satellite search-and-find missions to spot important military and industrial facilities. This picture was made at 9 P.M. over downtown Dallas from aircraft flying at altitude of 3,000 feet—giving much higher resolution than is obtainable from satellite altitudes. Traffic-flow pattern is visible at "A" (right center), and heat from truck exhaust can be seen near "B" (lower left). The dark line through the tall building visible at "C" (upper center) is an elevator shaft, whose temperature is different from that of the rest of the building.

Resolution of this picture of the lower tip of Manhattan Island, taken from an airplane flying at 8 miles' altitude, is believed to be comparable to that now being obtained from current recoverable reconnaissance satellites. Photo-analysts, using tools of their trade, can distinguish different types of automobiles on the bridges, and the wakes from ships are visible to the naked eye. This picture was taken using a Bournes/CAI camera with a 6-inch focal length.

Westinghouse Electric

Picture of Baltimore and adjoining harbor made with "side-looking radar," a special type that offers higher resolution than conventional radar. Side-looking radar may see limited use in new-generation reconnaissance satellites because of its ability to penetrate clouds, fog and darkness and to provide moderate resolution even from 100 miles' altitude. Resolution of this picture, made in 1958 with Westinghouse side-looking radar installed in an aircraft, is probably superior to what can be obtained from space at least for the near future. Much higher-resolution side-looking radar for aircraft is now in operational use.

through the grapevine, and NASA scientists had begun to think of using the new technology for civil applications, for what is now called an "earth-resources"-type survey.

Multi-spectral photographs taken from aircraft and/or satellites can be used to locate diseased or insect-infested crops and timber, to detect sources of water pollution, and to discover new mineral deposits—to cite but a few examples, all based on the "reflectance fingerprint" of the terrain. By 1966, NASA had launched an earth-resources surveying investigation, with strong support from other government agencies, such as the Agriculture and Interior Departments. The space agency now plans to launch two experimental unmanned satellites in 1972–73 to test these concepts and to conduct other earth-resources tests in its manned Skylab, slated for launch during the same period. Itek has been selected to supply a six-camera multi-spectral sensor for the Skylab, which will take four black/white photos, each in a different band, plus an infrared and a full-color photo of the same scene. The camera is expected to weigh approximately 180 pounds. Presumably it is an outgrowth of multi-spectral camera systems which Itek first began to produce around 1965 for use in a third generation of strategic reconnaissance satellite.

Another type of reconnaissance sensor probably made its debut on an experimental basis during the 1964–65 period as a prelude to being deployed on an operational basis in the third-generation satellites. This is the infrared scanner, which can produce photograph-like pictures during complete darkness and is especially useful for detecting important strategic facilities. The ability to "see" or detect objects in complete darkness results from the fact that every object, whose temperature is greater than "absolute zero" (−460° F.), generates and emits infrared radiation. Human beings, power plants, automobiles, even icebergs, emit such infrared radiation— but at a longer wavelength to which infrared film cannot respond. The wavelength at which an object emits its most intense infrared radiation reveals the temperature of the object.

The ability of an infrared sensor to "see" or detect an object depends upon the temperature differential between the object of interest and the surroundings. For example, it is relatively easy to spot a warm factory building if it is surrounded by snow-covered ground, or a warm ship in relatively cold waters. Concrete and asphalt roads become warm by absorbing sunlight during the day, but they lose this heat more quickly at night and therefore stand out clearly against the warmer ground on either side. Thus an infrared sensor is especially useful for detecting camouflaged targets, or objects whose significance might otherwise be overlooked. For example, a large power plant located in a remote part of the USSR, used to supply electricity to a secret underground facility but disguised or camouflaged to hide its importance, would be readily detected in a picture made by an infrared scanner and would prompt closer investigation of the area. The water used by a nuclear submarine to cool its reactors, and then exhausted in the submarine's wake, might be sufficiently warmer than the surrounding water to permit detection.

An infrared scanner employs special "detectors" which are sensitive to infrared radiation and convert it to an electrical signal of corresponding intensity. For some applications this electrical signal is used to produce a visual photograph-like image. (See Plate 14.) For strategic reconnaissance-satellite applications, the infrared scanner probably would be employed for large-area surveillance and would logically be carried aboard one of the radio-transmission-type spacecraft. Rather than use the infrared scanner's electrical signals to create a photographic film in the satellite, and then scan and convert the image back into an electrical signal for radio transmission, the original scanner signals probably are stored temporarily on magnetic tape.

The proliferation of new types of sensors was a boon to the photo-analyst, but it was a headache for those who had to plan and program the reconnaissance satellites to carry and use the appropriate sensor over each target area. Additionally, the growing variety of sensors that could now be carried into orbit produced a vast number of photos which had to be painstakingly analyzed, and much of the terrain photographed by the large-area surveillance satellites contained no targets of interest.

American astronauts who had gone into orbit in the tiny, cramped Mercury spacecraft starting in the spring of 1963 had reported being able to easily recognize landmarks with the naked eye. Some even claimed they could see individual houses and surface vehicles. This seemed to contradict terrestrial tests which indicated that from an altitude of 100 miles, the smallest object that could be resolved by the human eye was approximately 200 feet in size. (Subsequent tests indicate that the zero-G condition in space may be partially responsible for the higher acuity.)

This suggested a potential solution to the problems of selecting targets and appropriate sensors in reconnaissance satellites. Human observers in orbit, equipped with moderately powerful binoculars, could readily spot many objects of potential interest and could photograph them without delay. This would eliminate the many pictures of uninteresting terrain which now occupied the time of photo-analysts on the ground. Equally important, the astronauts could be trained to select the appropriate type of sensor for each type of target.

The Air Force had long hoped to find a space mission for military astronauts, and the experience of the Mercury astronauts provided ammunition to support its case. In addition to the important reconnaissance mission, military astronauts could warn of any surprise missile attack. (This mission will be covered in Chapter 18.) On January 23, 1965, the Pentagon announced that it had approved plans for industry studies of a Manned Orbiting Laboratory, which became known as MOL (pronounced "mole"). The Pentagon said that after completion of the studies, "a decision will be made whether to proceed with full-scale development." The announced purpose of MOL was "to determine the military man's potential usefulness in space." However, the program would "encompass development of technology to improve the capabilities for manned *or unmanned* operations of military significance," the Pentagon said. (Emphasis added.)

The upcoming launches by NASA of the larger two-man Gemini spacecraft, starting in the summer of 1965, offered the Defense Department the opportunity to run experiments to test further the usefulness of human astronauts for reconnaissance and picture-taking. Gemini-4, launched on June 3, was the first of the new spacecraft to remain aloft long enough for the astronauts to make a sizable number of pictures in addition to their primary task of checking out the new craft. The astronauts carried a slightly modified version of the Swedish Hasselblad camera, used by more affluent amateurs, with a standard 3-inch (80-mm.) lens. Despite the short focal-length lens, the several hundred color photos taken by the Gemini-4 astronauts showed remarkable detail.

Gemini-5, launched on August 21, 1965, carried two light-weight telephoto lenses, in addition to the standard 3-inch lens, which NASA announced would be used for "Defense Department-sponsored experiments." One of these was a 10-inch (250-mm.) lens, while the other was an optical giant, with a 48-inch (1,270-mm.) focal length, which was ingeniously squeezed into an overall length of 8 inches—small enough to be carried in the cramped Gemini spacecraft. The lens, made by a company named Questar, was originally built for use in small telescopes for amateur astronomers rather than for spaceborne photography.

Only a couple of the photos taken with the two telephoto lenses were ever released by the Defense Department, and these were intentionally "fuzzed up" to degrade their ground resolution. Despite this, one of the photos taken with the Questar lens over Love Field, the downtown Dallas airport, from an altitude of more than 100 miles, clearly shows individual runways, taxi-strips and buildings. (See Plate 11.) Another photo taken with the Questar lens of downtown Dallas was published in the October 11, 1965, issue of *Aviation Week & Space Technology*, with a caption that read in part: "Photo interpreters think they can identify the triple underpass near which President Kennedy was assassinated, at the convergence of Commerce St. and the river."

All of the color photos taken by the Gemini-5 astronauts with the 3-inch lens and Hasselblad camera were released,* and they were even more impressive than those from earlier

* Officially, NASA released all of the color photos, but three of the pictures which were taken while the spacecraft was passing over Cuba allegedly were so badly underexposed that the negatives turned out completely black. This is especially strange since all of the other photos taken by the astronauts have excellent exposure.

missions. One taken as the spacecraft passed over Cape Canaveral clearly shows roads, buildings and launch pads even when viewed with the naked eye. (See Plate 10.)

The Gemini-5 astronauts were not yet down from orbit on August 25 when the Soviet military newspaper *Red Star* charged that astronauts L. Gordon Cooper, Jr., and Charles Conrad, Jr., were "spying" on Communist countries with powerful cameras "that make possible detailed pictures of cities, railways, ports, ships, etc." Clearly, the author of the *Red Star* article had seen the NASA release which made no attempt to hide the Defense Department photographic experiments and he recognized the significance of a telephoto lens with a 48-inch focal length.

The same day that *Red Star* made its charges, President Johnson announced that he had approved the start of development of the Manned Orbiting Laboratory. Asked about the Soviet accusations, the President said he had invited the Russians to send representatives to the next Gemini launch. (They would not accept, for this would seem to obligate the Soviets to invite American observers to one of their own manned spacecraft launches, which remain, as of this writing, a very closed affair.)

Gemini astronauts continued to conduct photographic experiments for the Defense Department in subsequent missions until the program ended with the Gemini-12 flight in mid-November 1966. Included were tests of new types of photographic film, one an ultra-sensitive type with an ASA speed of 6,000. (By way of comparison, ordinary color film has an ASA speed of 25 and regular black/white film normally used by amateurs has an ASA speed of 400 or lower.)

By late 1965 the U.S. was getting ready to introduce a third generation of reconnaissance satellite, and studies were under way for still another, more advanced design which could become operational by the early 1970s. The U.S. also had decided to spend $1.5 billion to develop and test the MOL, a potential competitor to the fourth-generation unmanned reconnaissance satellites. Spaceborne reconnaissance had paid off handsomely and the U.S. was prepared to hedge its bets.

16

SOVIET RECONNAISSANCE SATELLITES

Analysis of the Soviet spaceborne reconnaissance effort shows significant differences between the approaches employed by the two countries, reflecting their different needs and capabilities. Because the U.S. openly publishes information on the general location of its missile bases and military airfields (believing that such information cannot be concealed in an open society), the Russians did not face as difficult a search-and-find mission as confronted the U.S.

Initially, the Soviets would want to make a general photographic surveillance to satisfy themselves that published U.S. information was correct and that there were no unreported strategic facilities. This would be easier for the Soviets than for the U.S. because the continental U.S. land-mass is only one-third the size of the Soviet Union. Once this had been done, the Soviets would need to pinpoint the precise location of each potential target so that its exact bearing and distance could be set into the guidance system of Russian ICBMs before launch. The Russians also would need to pinpoint the location of potential targets in West Europe for their large arsenal of shorter-range missiles.

The Russians would also want to monitor the deployment

of America's large, far-ranging Navy, especially the movements of Polaris submarines—if that were possible. Later, as Red China began to emerge as a potential military adversary in the 1960s, Russia would need to conduct search-and-find reconnaissance missions over this large Asian land-mass to try to penetrate the "Bamboo Curtain."

The orbital-recovery technique, which the Russians had first developed to return canine, and later human, astronauts from space, enabled them to rapidly develop a recoverable reconnaissance-satellite capability. Circumstantial evidence suggests that the Soviets resorted to recoverable camera/film-pack techniques for close inspection and precision mapping (as did the U.S.) and for some types of search-and-find missions. It appears that the Russians use radio-transmission-type reconnaissance satellites only for low-resolution oceanic-surveillance and cloud-cover photography in support of their strategic reconnaissance satellites.

It seems safe to assume that Russia's high-altitude aerial-camera technology was not nearly as advanced as that of the U.S. when their spaceborne reconnaissance effort began. The U.S. had pioneered in high-altitude aircraft for bombing and reconnaissance, beginning even before World War II. The Russians did not begin to develop and build high-altitude aircraft until the late 1940s, a decade later.

Analysis of Soviet recoverable satellites, launched under the multi-function Cosmos program, shows that they began to acquire a near-clocklike regularity during the 1964–65 period, about the same time as the U.S. program. There is no way of knowing what quality of photos was obtained, or how many of the satellites brought down from orbit were successfully retrieved on the ground. So far as is known, the Soviets have not developed a midair-recovery technique because of the huge size and 10,000-pound weight of the spacecraft which is returned in its entirety.

The first three recoverable Cosmos satellites launched in 1962 remained aloft for only four days, while the fourth stayed in orbit for eight days. During 1963 the Russians launched seven of this type, and most stayed in orbit for at least eight days before being returned. All of the recoverable satellites were launched into a 65-degree-inclined orbit.

If there was an imprecise reentry, resulting in an over-shoot of a few hundred miles, the spacecraft would parachute down into uninhabited northern regions of the Soviet Union. During the winter months, with such regions under heavy snow cover, retrieval would be difficult if not impossible. The fact that the Russians did not launch any recoverable satellites during the winter months for the first several years suggests that they had not yet mastered precise reentry and recovery.

During 1964, following a three-month winter launch-gap, the Soviets orbited 12 of the recoverable satellites, an increase of roughly 70 percent over the previous year, as the following tabulation shows. The time of launch of the Soviet satellites was slightly later in the afternoon than for U.S. spacecraft, generally between 1 P.M. and 4 P.M., Tyuratam time.

SOVIET RECOVERABLE SATELLITES ORBITED IN 1964

Cosmos #	Launched	Orbit-Life	Perigee	Apogee
28	Apr. 4	8 days	130 mi.	245 mi.
29	Apr. 25	8	127	192
30	May 18	8	128	238
32	June 10	8	130	207
33	June 23	8	130	182
34	July 1	8	127	224
35	July 15	8	135	167
37	Aug. 14	8	127	186
45	Sept. 13	5	128	203
46	Sept. 24	8	134	168
48	Oct. 14	6	126	183
50*	Oct. 28	8	122	150

* Exploded on 8th day, possibly due to retrorocket malfunction.

Note: The reader may be able to identify future Cosmos satellite launches involving recoverable-type satellites if his local newspaper carries the brief Russian announcements of new satellites orbited. If the satellite perigee is 220 kilometers or less and the inclination is 65 degrees or more, the satellite probably is a recoverable-reconnaissance type.

The perigee altitude of the recoverable Russian satellites is nearly one-third higher than for their American equivalents —which would result in a corresponding reduction in photographic resolution. One possible explanation is that the Russian satellites, believed to be large-diameter circular-shaped craft, would exhibit much higher drag than the long, narrow Agena and therefore would require a higher perigee altitude to achieve the desired lifetime.

Beginning in 1965, the Russians launched their recoverable satellites through the winter months, indicating increased mastery of reentry technology. A total of 17 of this type of satellite were orbited, almost a 50 percent increase over the number launched the previous year.

SOVIET RECOVERABLE SATELLITES ORBITED IN 1965

Cosmos #	Launched	Orbit-Life	Perigee	Apogee
52	Jan. 11	8 days	127 mi.	189 mi.
59	Mar. 7	8	130	211
64	Mar. 25	8	128	168
65	Apr. 17	8	130	213
66	May 7	8	122	181
67	May 25	8	129	217
68	June 15	8	127	208
69	June 25	8	131	206
77	Aug. 3	8	124	181
78	Aug. 14	8	128	204
79	Aug. 25	8	131	223
85	Sept. 9	8	132	198
91	Sept. 23	8	132	213
92	Oct. 16	8	132	219
94	Oct. 28	8	131	182
98	Nov. 27	8	134	354
99	Dec. 10	8	124	199

It is curious that the Russians should orbit 17 recoverable satellites in 1965, more than twice the number launched by the U.S. that year, to perform reconnaissance over a smaller land-mass that had no Iron Curtain secrecy. One possible explanation is that the Soviets were using recoverable satellites both for search-and-find and close-inspection-type missions. Another is that Russian camera systems were not yet as reliable as their American counterparts and therefore required more launches to produce comparable numbers of useful photos.

All of the recoverable Russian satellites had been launched, so far, from the large Tyuratam facility because the Kapustin

Yar site could only handle smaller satellites that could be orbited with an IRBM first stage. As the pace of reconnaissance-satellite launches increased and the scope of the Soviet space program expanded, a "traffic jam" was developing at Tyuratam. Brief gaps in the launch of recoverable reconnaissance satellites would occur, for instance, at times when Tyuratam was needed to launch manned spacecraft.

Also, Tyuratam's geographic location, originally selected to test ICBMs, was not ideal for reconnaissance-satellite orbits. Any rocket booster can orbit the maximum possible payload if it is fired due east, to take advantage of the earth's own eastward rotational velocity. If a satellite is launched into a near-polar orbit, which has advantages for reconnaissance missions, approximately 20 percent of the payload weight is sacrificed, as compared with a due-east launch. But when a satellite is launched due east, for maximum payload, it will not orbit at a latitude higher than the latitude of its launch site. For example, a due-east launch from Tyuratam, at 45 degrees north latitude, means the satellite will not pass over much of West Europe or the northern portions of the U.S. and Canada. The launch vehicle can be fired to the northeast to achieve higher latitude coverage, as the Russians were doing to obtain a 65-degree inclination, but this reduces the payload.

Even if the Soviets were prepared to make the payload sacrifice, it was not feasible to launch a satellite into a near-polar orbit from Tyuratam because of the risk that the first-stage rocket might fall on inhabited areas in central Russia. A near-polar orbit would also be desirable for the civil meteorological satellites, similar to Tiros, which the USSR was then developing. Clearly the Russians needed a new launch site, and by 1965 construction was under way near the town of Plesetsk, approximately 600 miles due north of Moscow. From Plesetsk, at a latitude of 65 degrees, a due-east launch with maximum possible payload would take a Russian satellite far enough north to cover most of the areas of interest in North America and West Europe. Furthermore, when desired, a satellite could be lofted into a near-polar orbit without risk to inhabited areas.

The first recoverable satellite to be launched from the new site, Cosmos-112, went into orbit on March 17, 1966, and remained aloft for the customary eight days. Five more of this type were launched from Plesetsk during the rest of 1966. Four of the six went into 72–73 degree-inclined orbits, taking the Soviet satellites to higher latitudes than they had previously been able to reconnoiter. Now, for example, Russian satellites could photograph early-warning radar networks in northern parts of Canada, Alaska and Greenland. Of the 21 recoverable satellites launched by the USSR in 1966, 15 still came from Tyuratam—but at least the traffic jam had been eased somewhat. (Satellites orbited from Plesetsk, like those from Tyuratam, usually are launched in the afternoon, between 1 P.M. and 4 P.M. local time.)

(The Russians have never announced or officially admitted the existence of the new launch site, nor has the U.S. government. The honor of "discovering" and first making public the new Plesetsk facility goes to the students of England's Kettering Grammar School and its science teacher, Geoffrey E. Perry, who first interested the boys in tracking Russian satellites and plotting their orbits, using hand-me-down equipment. When the students tried to plot the orbit of Cosmos-112, they were perplexed because the satellite did not seem to have originated either at Tyuratam or Kapustin Yar. When four more Cosmos satellites launched during the next few months showed the same puzzling characteristics, and all seemed to have originated from the same launch point near Plesetsk, the Kettering students decided that the Russians must have built a new launch facility, which indeed they had.)

However, the Russians presently are not able to launch a satellite into a "sun-synchronous orbit" with a 97-degree inclination, which is especially useful for meteorological and long-duration reconnaissance satellites. The probable reason is that they would need tracking-guidance stations located in Norway or Sweden for the required launch to the northwest. During 1967 the Russians launched 23 recoverable satellites, two more than during the previous year. Of this total, 14 were orbited from Plesetsk and only 9 from Tyuratam, further easing the work load at the older facility.

There is another type of satellite launched under the Cosmos label which goes into moderately low perigees and remains aloft for several months before it decays. The Soviets claim that these are scientific satellites, designed to measure natural (Van Allen) radiation and other space phenomena. At first, Western observers (not privy to intelligence information) accepted this explanation. But as the number of this class of Cosmos placed in orbit rose from three in 1963 to thirteen in 1968, the Soviet claim has lost credibility. There simply is not that much of scientific interest to measure at such low altitude to justify so many satellites.

My own analysis suggests that this class of Cosmos satellite is a radio-transmission type similar to the original Samos, with moderate resolution, which is used both for oceanic surveillance and to obtain cloud-cover pictures for use in programming Soviet recoverable photo-reconnaissance satellites. Moderate resolution would be sufficient to spot surface ships with long wakes, as shown by American astronauts who were able to see vessels with the naked eye. My analysis suggests that this radio-transmission-type satellite is an adaptation of the Lunik-3 spacecraft, which the Soviets used in October 1959 to take and send back photos of the backside of the moon.

The Lunik-3 spacecraft weighed 614 pounds, according to Russian figures, and measured about 4 feet in diameter and slightly less than 5 feet in length. These are similar to dimensional estimates for the "mystery class" of Cosmos satellites made by Britain's Royal Aircraft Establishment based on its radar measurements and satellite orbital-decay time. The RAE also "guestimates" the weight of this Cosmos type at approximately 900 pounds. The spacecraft were first launched from Kapustin Yar using an IRBM as a first stage, so the 900-pound estimate appears reasonable.

If the Russians had decided in 1959 or early 1960 to develop a reconnaissance-satellite capability patterned after the American Samos, whose existence was by that time known to them, it would have been logical to adapt the Lunik-3 spacecraft for this mission. Like Samos, Lunik-3 had carried a camera whose film was developed on board and was scanned with an electric-eye to convert the image into signals suitable for transmission by radio-link. The original design could have been modified to carry a larger film supply, more batteries to permit radio transmission during darkness and a higher-resolution lens without raising the original 614-pound spacecraft weight beyond the 900 pounds that could be orbited by an IRBM first stage.

It seems more than mere coincidence that the early ener-getic Soviet program to explore and photograph the lunar surface with unmanned satellites suddenly halted after the Lunik-5 mission in October 1959, and did not resume for more than three years—in April 1963. The first of the "mystery class" of Cosmos satellites was launched on March 16, 1962.

My analysis suggests that this type of nonrecoverable Cosmos satellite is used for *two* different types of missions. Some of the spacecraft are lofted into perigee altitudes of roughly 130 miles, approximately the same altitude used for Russian recoverable reconnaissance satellites, which can provide reasonably good photographic resolution. But other satellites of this general class go into a higher perigee of approximately 175 miles, which would degrade resolution significantly and reduce the useful payload for no seeming purpose if these were intended for surface-target reconnaissance.

However, if these higher-perigee Cosmos are intended to make cloud-cover photos, then the greater altitude would be an operational advantage because fewer pictures would have to be taken and high resolution is not needed. Unlike the U.S., which began to operate Tiros cloud-cover weather satellites in early 1960, before the first photo-reconnaissance satellites were introduced, Russia did not begin to orbit "civil" weather satellites until May 11, 1966, with the launch of Cosmos-118. This was more than four years after Russia had begun to orbit photo satellites, and without some sort of cloud-cover information, Soviet reconnaissance satellites would have wasted much of their precious film.

Circumstantial evidence to confirm this cloud-mapping mission for some of this class of nonrecoverable Cosmos satellite can be found in the fact that shortly after the Russians began to launch their recoverable reconnaissance satellites into more inclined orbits from Plesetsk in 1966, they began to launch many of the nonrecoverable Cosmos satellites believed to have a cloud-cover sensing mission from the new facility also. If increasing numbers of recoverable photo satellites were to operate at higher latitudes, Russian programmers would need cloud-cover pictures of these northern regions, and satellites launched from Kapustin Yar were limited to latitudes below 50 degrees.

This shift is apparent in the following tabulation of the nonrecoverable Cosmos satellites thought to have a cloud-cover reporting mission. Satellites launched from Plesetsk are identified by "(P)."

Cosmos #	Launched	Decayed	Perigee	Apogee
1965:				
76	July 23	Mar. 16, 1966	162 mi.	329 mi.
101	Dec. 21	July 12, 1966	162	342
1966:				
106	Jan. 25	Nov. 14	174	350
116	Apr. 26	Dec. 3	183	297
123	July 8	Dec. 10	163	329
135	Dec. 12	Apr. 12, 1967	161	411
1967:				
148 (P)	Mar. 16	May 7	171	271
152 (P)	Mar. 25	Aug. 5	176	318
163	June 5	Oct. 11	162	383
166	June 16	Oct. 25	176	359
173 (P)	Aug. 24	Dec. 17	174	328
191 (P)	Nov. 21	Mar. 2, 1968	175	322
1968:				
204 (P)	Mar. 5	Mar. 2, 1969	175	542
222 (P)	May 30	Oct. 11	172	328
225	June 12	Nov. 2	160	329
230	July 5	Nov. 2	180	360
242 (P)	Sept. 20	Nov. 13	174	273
245 (P)	Oct. 3	Jan. 15, 1969	175	316
257 (P)	Dec. 3	Mar. 5, 1969	175	292
262	Dec. 26	July 8, 1969	163	508

Since early 1969, practically all Cosmos satellites of this type have been launched from Plesetsk.

If the other small nonrecoverable Cosmos satellites with lower perigee altitudes are used for oceanic reconnaissance, as I suspect, they could still be launched mostly from Kapustin Yar because their 48-degree inclination would take them over the more heavily traveled shipping lanes. Periodically the

Russians could launch satellites of this type from Plesetsk into a near-polar orbit for reconnaissance of arctic shipping lanes during the summer months when they become usable. This is exactly what an analysis of these Soviet satellites shows. Beginning June 12, 1967, the Russians launched two satellites of this type into 82-degree-inclined orbits from Plesetsk, but the rest of the launches continued to be made from Kapustin Yar.

Cosmos #	Launched	Decayed	Perigee	Apogee
1965:				
53	Jan. 30	Aug. 12, 1966	141 mi.	741 mi.
70	July 2	Dec. 18, 1966	142	717
93	Oct. 19	Jan. 3, 1966	137	324
95*	Nov. 4	Jan. 18, 1966	129	324
1966:				
108	Feb. 11	Nov. 21	141	537
119	May 24	Nov. 30	136	811
137	Dec. 21	Nov. 23, 1967	143	1,069
1967:				
142	Feb. 14	July 6	133	846
145	Mar. 3	Mar. 8, 1968	137	1,327
165 (P)	June 12	Jan. 15, 1968	131	958
176 (P)	Sept. 12	Mar. 3, 1968	128	982
196	Dec. 19	July 7, 1968	140	551
197	Dec. 27	Jan. 30, 1968	137	314
1968:				
202	Feb. 20	Mar. 24	137	312
211 (P)	Apr. 9	Nov. 10	130	978
219	Apr. 26	Mar. 2, 1969	138	1,100
221	May 24	Aug. 31, 1969	137	1,310
233 (P)	July 18	Feb. 7, 1969	130	960
259	Dec. 14	May 5, 1969	136	841

*Exploded in orbit.

The Russians had used the new Plesetsk site to launch *recoverable* reconnaissance satellites into 72–73-degree-inclined orbits since early 1966, but it was almost two years later before they achieved a near-polar orbit with Cosmos-210, launched into an 81-degree inclination on April 3, 1968. This was followed by another recoverable satellite, Cosmos-214, into an 81-degree inclined orbit on April 18. These higher inclinations, which require more powerful rockets, as well as the introduction of longer-endurance recoverable satellites at about the same time, suggest an improved model of the Soviet's large booster.

The first long-endurance recoverable satellite, Cosmos-208, was launched on March 21, 1968, from Tyuratam into a conventional 65-degree inclination and it remained aloft for twelve days. Another twelve-day recoverable satellite, Cosmos-228, was launched three months later, on June 21; and still another, Cosmos-243, on September 23. Thereafter, the Russians would launch the twelve-day recoverable spacecraft at irregular two-to-four-month intervals, interspersed with the older eight-day-mission spacecraft.

During 1969 the Russians launched 33 of their recoverable reconnaissance satellites, while the U.S. launched a total of only 12 of the radio and recoverable types. The Soviets kept a recoverable type of reconnaissance satellite in orbit for 233 days out of the year, or 64 percent of the time. And for forty-six days out of the year, there were two Russian photo satellites in orbit simultaneously. The reason for this accelerated pace is not difficult to explain, for it was in 1969 that the long-festering relations between the USSR and Red China broke out in bloody pitched battles along their borders in eastern Siberia.

Prior to 1969 the Russians had been launching a recoverable satellite approximately every two weeks. But beginning a week before the bitter fighting over Damansky Island on the Ussuri River in the USSR's Eastern Maritime Province, the Russians began to launch at double the usual rate! Between February 25 and April 23, Russia launched a total of 10 recoverable reconnaissance satellites. Clearly, Soviet leaders were anxious to know what was taking place behind the Bamboo Curtain.

In May, the launch pace settled back to the regular rate as the two sides hurled recriminations at each other over the two Damansky Island clashes. Then, on June 10, armed conflict broke out again near the western Mongolian border. Once again the pace of Soviet satellite launches accelerated. Between June 15 and July 22, the Russians put five recoverable satellites into orbit. During the first two weeks of August, no Russian reconnaissance satellites of the recoverable type, whose high resolution would be needed to spot troop movements, were in orbit. But on August 13 another fight occurred on the Mongolian border, and the Soviet space-reconnaissance effort once again shifted into high gear. A recoverable satellite was orbited on August 16, another on August 19, another on the 29th and still another on September 2. The Russians kept up their reconnaissance guard by launching three more of this type during the rest of September.

On September 11, Soviet Premier Alexi Kosygin journeyed to Peking for the funeral of North Vietnamese leader Ho Chi Minh and had discussions with Red Chinese Premier Chou En-lai. This set the stage for lower-level, more formal talks that began on October 20 in an effort to resolve the border issues. Following Kosygin's visit to Peking, and the relaxation of tensions, the pace of Russian photo-satellite launches settled back to the normal average of roughly one every two weeks.

The Iron Curtain and Bamboo Curtain hinder the West from knowing the current state of relations between the two giants of the Communist world. Hence the launch rate of Soviet reconnaissance satellites may provide a barometer which bears watching.

In an article written for *Aviation Week & Space Technology,* which appeared in the April 6, 1970, issue, I pointed out this apparent correlation between the pace of Soviet reconnaissance-satellite launches and Sino-Soviet relations. The article noted: "Red China's lack of satellite reconnaissance, which keeps it in the dark about possible Soviet military actions, prompts some observers to expect that Red China will seek to develop reconnaissance spacecraft as soon as its rocket booster progress permits." Less than three weeks later, on April 24, Red China launched its first satellite.

Red China's first satellite was small and unsophisticated. But so were Sputnik-1 and America's first satellites, launched in the late 1950s. Yet within several years, both Russia and the U.S. were orbiting reconnaissance satellites. Before the decade is out, perhaps even by 1975, Soviet and American reconnaissance spaceships that pass in the night may be joined by a new fellow traveler.

17

THIRD-GENERATION
U.S. RECONNAISSANCE SATELLITES

The debut of a new generation of U.S. recoverable reconnaissance satellites in mid-1966 is revealed by the introduction on
July 29 of a more powerful launch vehicle to orbit such spacecraft—the Titan-3B/Agena-D, an adaptation of the Titan-2
ICBM. The new booster could loft more than 6,000 pounds
of useful payload into reconnaissance-type orbit, or roughly
1,500 pounds more than the latest version of the previously
used Atlas/Agena-D. Some of this added capacity could be
used to carry more consumables, such as film and electric
batteries, to extend satellite endurance.

The earliest recoverable satellites had been able to carry
enough consumables to remain in orbit for only 4 to 5 days,
but by 1966 this had been increased to an average of 7 days.
The first of the third-generation recoverable satellites, launched
on July 29, remained in orbit for 8 days, and the second,
launched on September 28, stayed aloft for 9 days. The third,
fired on December 14, remained in orbit for 10 days. By 1968,
many of the recoverable spacecraft were remaining aloft for
15 days or longer.

As a result, the U.S. was obtaining more total orbital time
with significantly fewer recoverable satellite launches. During

1968 it required only eight satellites to achieve 99 days in orbit, a third fewer spacecraft than the U.S. had needed in 1966 to obtain 86 days in space. During 1969, the U.S. launched only six of the new-generation satellites, with each averaging 14 days in orbit—twice the 1966 average. In 1970 the U.S. obtained a total of 85 orbit-days from only five recoverable satellites, compared with 82 orbit-days from six spacecraft the previous year. (In 1970 the number of Soviet recoverable satellites declined for the first time, with a total of 29 spacecraft orbited, compared with 33 in 1969. Nearly half of the Soviet recoverable satellites orbited in 1970 remained aloft for 12 to 13 days, whereas previously most of the Russian spacecraft had only 8 days' endurance.)

Beyond longer endurance, the new-generation U.S. spacecraft also could carry more sophisticated sensors to cope with the consequences of the new family of Soviet ICBMs that were being deployed in underground silos. In 1963, Defense Secretary McNamara had predicted that the Russians would harden their missiles, and in January 1964 he told Congress that the U.S. had evidence that this was being done. This might make it more difficult to locate and count Russian missiles, if the Soviets resorted to camouflage and/or overcast skies to hide their efforts.

More important was the effect of Soviet hardened missiles on the then existing U.S. strategy, which the Pentagon called "damage limiting." If thermonuclear war should break out, the Pentagon strategy called for American ICBMs to first attack known Soviet missile sites in the hope of catching any weapons that had not yet been fired, either by intent or because of malfunction. This could reduce the number of Soviet missiles that were launched against the U.S., hence the term "damage limiting."

Back in the early 1960s, when the Soviets had only the giant SS-6 ICBMs and these were deployed in above-ground facilities, the Russian missiles could be disabled if an American warhead exploded even several miles away. But with underground silos, a direct hit or near miss would be needed to knock out Soviet ICBMs. For example, if Russian silos could withstand a blast pressure of 300 pounds per square inch, the design objective for U.S. silos, then an American

Minuteman with its one-megaton warhead would have only a 10 percent chance of knocking out a Russian missile if the warhead hit one mile from the silo. But the odds would go up to nearly 80 percent if the Minuteman could hit within one-quarter mile of its target.

(To better appreciate what a one-quarter-mile accuracy means for an ICBM traveling a 6,250-mile trajectory, imagine you were blindfolded and told to try to walk a perfectly straight line for precisely one mile to where you would find a tiny, precious gem. If you carefully paced off an estimated mile, removed the blindfold and looked down to see that the gem was *less than three inches* away from the tip of your shoe, you would have navigated with an accuracy equivalent to one-quarter mile for an ICBM.)

"Damage limiting" required that the accuracy of ICBM guidance be improved, and this was being done in the Minuteman-2 version. Equally important, the precise location of each Soviet missile silo would need to be known much more accurately than for a target such as a major city. This kind of information can be obtained from satellite photos, but it requires a different kind of camera, called a mapping or metric type, which has extremely high "geometric fidelity," rather than especially high resolution. The mapping camera is used only after targets have been spotted and identified by higher-resolution cameras.

Pictures from mapping cameras are carefully analyzed to measure the distance of the target to some landmark along one edge of the photo frame. The measurement is then extended, using adjacent frames until it can be "anchored" to a landmark whose precise position has previously been established. This basic landmark may be located outside the USSR so that its precise position can be established by means of radio or geodetic satellite techniques. Once the USSR had been accurately charted, the location of a newly discovered target could be referenced to a landmark within the USSR whose position had previously been established. Mapping cameras normally use a moderate 6-inch focal-length lens with a reasonably wide field of view to minimize the number of individual frames that must be used to relate a new target to a known landmark.

Ideally, the lens of a mapping camera should look straight down because any obliqueness in viewing angle will introduce geometric distortion. The Agena spacecraft itself, from the earliest days, had been equipped with horizon sensors and a stabilization system to orient its cameras toward the earth. But to achieve the extreme accuracy needed for mapping, it seems likely that the metric camera would simultaneously photograph several stars in the celestial sphere. Later, these star photos would be used to determine the mapping camera's precise vertical angle and to introduce any necessary corrections.

The "damage-limiting" strategy, when applied against hardened Soviet missiles, demands such high accuracy that the film would have to be returned to earth rather than attempt to scan-convert the pictures and transmit them by radio. Furthermore, the temperature of the film would have to be accurately maintained during the mission because a dimensional change of even a few ten-thousandths of an inch would introduce serious errors.

Mapping cameras had been developed some years earlier for aircraft use, and it seems likely that they were adapted and tried in satellites launched during the early 1960s. But the need for ultra-accurate data on the location of Soviet missiles did not emerge until around 1963. If development of a mapping camera specifically for the satellite mission had been started at that time, the first thoroughly tested models would have been ready for operational use around mid-1966, in the third-generation satellites.

Another new sensor introduced operationally in the third-generation recoverable spacecraft was the multi-spectral camera, discussed in Chapter 15. Because of space and weight constraints, the camera probably operated in four bands or at most in six, and was built by Itek Corp., which pioneered in this technology. (A recent Itek Stockholders Report refers cryptically to "continued active participation in strategic reconnaissance.")

The introduction of multi-spectral cameras in reconnaissance satellites should be marked by a change in the time of spacecraft launch from the usual high-noon sun angle. The reason is that the colors reflected from ground objects vary with time of day, and season, so photo interpreters would want pictures taken of unfamiliar objects at different times. Analysis shows that beginning in 1966–67 the launch time for some recoverable satellites was advanced to the late morning, and by 1970 some were being fired as early as 7 A.M., local standard time.

A new generation of radio-transmission-type satellite also was introduced in the summer of 1966, as indicated by the debut of a more powerful booster on August 9. The new launcher was the Long-Tank-Thrust-Augmented-Thor/Agena-D, in which the length of the Thor first stage had been increased nearly 20 percent (to 71 feet) to accommodate more fuel and permit a longer thrusting period. The new LTTAT/Agena-D could orbit more than 2,600 pounds of useful payload, or roughly 20 percent more than the previous TAT/Agena-D. (For some subsequent missions, this increased payload would be used to carry a second "piggy-back" satellite for electromagnetic reconnaissance—a function discussed in Chapter 19.)

This added payload enabled the third-generation radio-transmission satellite to carry not only a higher-resolution camera system but also an infrared scanner, which can make reconnaissance pictures during night-time passes, as described in Chapter 15. Although the infrared-scanner pictures would have somewhat poorer resolution than ordinary camera photos made in daylight hours, the infrared scanner would be extremely useful in spotting "warm" targets of possible strategic import. (See Plate 14.) These targets could later be photographed by using high-resolution cameras in the recoverable-type satellites.

The infrared-scanner signals presumably are temporarily stored on magnetic tape in the satellite and transmitted down to terrestrial stations in the same manner as visible-light photos. The only difference would be that ordinary photos taken over the USSR and Red China during daylight would be transmitted to stations on the opposite side of the globe during darkness, while infrared pictures made during darkness would be sent back during daylight hours.

The infrared technology used in the satellite scanners

originally was developed by scientists at the University of Michigan. The spaceborne equipment used is believed to have been manufactured by HRB-Singer and by Texas Instruments Inc.

The third-generation radio-reconnaissance satellite also was outfitted with a newly developed "Space-Ground Link System" (SGLS) which enabled the spacecraft to transmit pictures at a much faster rate than before. This meant that several times as many reconnaissance pictures could be sent to a ground or shipboard station during each pass.

The nine-month lull that followed the first LTTAT/Agena mission in August 1966, before there was another on May 9, 1967, suggests that there were some problems with the spacecraft design. During this nine-month period, the USAF continued to launch radio-transmission satellites using the older TAT/Agena, but these would have been the lighter satellites of the earlier vintage. Following the May 1967 resumption of LTTAT/Agena launches, all radio-transmission satellites would be orbited using the new booster. During that year, a total of six new-generation spacecraft were orbited and each remained aloft for approximately four weeks.

The expanded capability of the third-generation radio-transmission satellite introduced in 1966–67, with its increased supply of film, electric power and other consumables, has made it possible for the U.S. to gradually reduce the number of spacecraft of this type launched each year for search-and-find missions. Whereas the U.S. orbited 13 satellites of this type in 1965, the number was cut to eight in 1968, six the following year and only four in 1970.

It will be recalled that two different techniques, described in Chapter 15, had been devised to relay received pictures from shipboard or remote ground stations back to Washington. One was to re-transmit the tape-recorded signals to a courier aircraft, flying in the vicinity of the shipboard or remote ground station; the other was physical recovery of the tapes by a low-flying aircraft equipped with a suspended trapeze-like hook. (For less remote ground stations, such as the one in New Hampshire, more conventional transportation was used.) Beginning in late 1967 or early 1968, a speedier means of relaying the photos to Washington is believed to have come into use. This is the use of communications satellites.

The first veiled hint of the new technique came in late 1967, when the Pentagon's chief scientist, Dr. John S. Foster, Jr., revealed that military communications satellites were being used to relay high-resolution photos, taken by reconnaissance *aircraft*, from South Vietnam to Washington. The new technique carried the code name of Compass Link. Later the Pentagon issued a guarded announcement which mentioned that these aerial photographs were converted to electrical signals for satellite transmission by means of an electro-optical scanner made by CBS Laboratories and signal-processing equipment supplied by Philco-Ford. *These are the same companies that produce similar equipment for use in the radio-transmission-type reconnaissance satellites.*

If aerial photos with high resolution can be transmitted from South Vietnam to Washington by military communications satellite after having been converted into electrical signals by CBS and Philco-Ford equipment, then reconnaissance-satellite photos in the same electrical format, as received by ground and shipboard terminals, could also be relayed directly to Washington by the same procedure. Two terminals near Washington presumably are used to receive the pictures. One, operated by the Navy, is located near Annapolis, about a 45-minute drive from Washington; the other, situated at Andrews AF Base, Md., is about 25 minutes from the Pentagon by car—even less by helicopter.

(The film-pack from recoverable reconnaissance satellites normally is retrieved in the Pacific and returned to Hickam AF Base, Hawaii. There it is transferred to an Air Force KC-135 jet transport for the flight to Andrews AFB, near Washington.)

When development was first begun on the Manned Orbiting Laboratory (MOL) in mid-1965 as a contender for future space reconnaissance and early-warning missions, it had been estimated that $1.5 billion would be needed to develop, build and launch the first five spacecraft. The craft was expected to weigh about 25,000 pounds, of which approximately 5,000 pounds would be reconnaissance-equipment payload; the rest would be structure and life-support system for the crew. The 5,000 pounds were less reconnaissance-system weight than was already being carried in third-generation unmanned recoverable satellites. But much of this weight was for complex electronic-control equipment which would not be needed if there were humans aboard.

Original plans called for the MOL to remain in orbit for about a month, returning its reconnaissance film and magnetic tapes by periodically ejecting recoverable capsules. After a month the crew would return to earth, using a modified Gemini spacecraft. Eventually, it was hoped, the MOL could have a long life, and replacement crews could be ferried up, by means of the modified Gemini which then would be used to return the retiring crew.

By the summer of 1966 the projected weight of the MOL had grown to more than 30,000 pounds, and this required the development of a more powerful launch vehicle in which two solid-propellant rockets would be strapped onto the Titan-3 frame. (This version was called the Titan-3M.) By the spring of 1967 the original launch date for the first MOL had slipped from 1968 to late 1969 or early 1970, and the estimated project cost had climbed from $1.5 billion to $2.2 billion. By the end of 1967 the date for the first launch had slipped again, now to late 1970 or early 1971. Some of this slippage was due to diversion of funds for the war in Vietnam.

By June 1969 the estimated cost for the first five MOLs had climbed to $3 billion, or twice the original estimate, and the new Nixon administration was facing a budget crunch due to rising Vietnam war costs. Reluctantly Defense Secretary Melvin R. Laird decided to cancel the MOL project. By this time, the design of a fourth generation of unmanned reconnaissance satellite was far enough along to indicate that it probably could perform most of the functions planned for MOL and could do so at lower total cost. While human judgment was extremely useful in space reconnaissance, the cost in terms of spacecraft payload to maintain human astronauts in the hostile environment of space resulted in a questionable trade-off.

• • •
The first of the fourth-generation unmanned reconnaissance satellites is being assembled and tested at the Lockheed
plant in Sunnyvale, Calif., as this is being written. It will be
a giant spacecraft, probably 50 feet long and 10 feet in diameter, weighing more than 25,000 pounds. It probably will carry
close to 20,000 pounds of useful reconnaissance payload, or
nearly twice the combined weight of the payloads of a radio-
transmission and recoverable-type satellite now in use. Big
Bird, as it is unofficially called, is to be launched by a giant
Titan-3D/Agena (with two strap-on solid rockets), which will
stand as high as a 17-story building.

First launch of Big Bird was scheduled for late 1970 or
early 1971. But rumors of delays in some of the camera systems indicate that the first launch may not come until mid-
1971. The new satellite is expected to be able to perform the
functions of both the radio-transmission and recoverable-type
spacecraft now in use. Big Bird is expected to have an orbital
life of at least several months and will probably carry several
recoverable capsules so that film-packs can be returned at
frequent intervals. The spacecraft will carry the most sophisticated and expensive camera ever designed for spaceborne use.
Built by Perkin-Elmer, it is expected to have a resolution
measured in inches. In all probability the satellite's payload
will be flexible and can be tailored to carry the types of sensors needed for specific intelligence-gathering missions.

There is speculation that Big Bird may carry, at least on
some missions, a new type of spaceborne sensor which can
penetrate heavy cloud cover that otherwise obscures ground
targets from the view of conventional cameras and infrared
scanners. This is a special type of radar. Conventional radar
can penetrate clouds easily, but it has poor resolution, especially at long distances. For example, a conventional radar
operating at a frequency of 10,000 megacycles, which would
provide moderately good ground resolution when used in an
airplane, would be worthless at an orbital altitude of 100
miles. Even if this conventional radar used a large 10-foot-
diameter antenna, from an altitude of 100 miles it could not
"see" objects smaller than roughly *one mile in size.*

A new type of radar, conceived in the early 1950s, showed
promise of providing much higher resolution when carried

aboard a moving vehicle—providing its antenna was aimed to the side rather than straight forward as with conventional radar. The theory of this "side-looking radar," as it is called, is too complex to consider here, but by the early 1960s it was demonstrating remarkable resolution when used in reconnaissance aircraft. Photos of terrain made with side-looking radar have such high resolution that to an inexperienced observer they might seem to have been made with a conventional camera. (See Plate 16.)

If a side-looking radar designed for aircraft use were placed aboard a satellite in orbit at 100 miles altitude, it could provide a ground resolution of approximately 100 feet. While this is vastly better than a conventional radar, still it would be useful only against relatively large targets, such as ships at sea. More recently, an improved "focused" type of side-looking radar has come under investigation. In theory, at least, it should be able to provide a ground resolution of a *few feet* from an altitude of 100 miles. If this performance can be achieved, it would be an invaluable spaceborne sensor for surveillance of areas which are usually under heavy cloud cover, such as the Arctic. It would be surprising if Big Bird were not used at least to test the feasibility of side-looking radar.*

Another new sensor which may find use aboard Big Bird is a television camera with a "zoom"-type telephoto lens which could transmit to human observers on the ground the view they would see if they were aboard the spacecraft. If an unfamiliar object of interest were spotted, a radio signal could be sent to the spacecraft to immediately photograph the object with an appropriate lens. Then, on the spacecraft's next pass over a ground or shipboard station, the picture could be sent to earth. This would greatly speed up the process of detecting and photographing objects or activities of interest, especially transient ones, such as the movement of mobile ballistic missiles mounted on trailers.

To achieve such "real-time" reconnaissance capability will

---

* The use of side-looking radar is also under consideration for use in an ocean-surveillance satellite, under study by the U.S. Navy, intended to monitor the USSR's fast-growing naval operations. This program currently carries the code number 749.

require the use of a special communications satellite to relay the TV picture from the low-altitude reconnaissance spacecraft to ground stations in Guam, Australia or ships based in the Indian Ocean. From these terrestrial terminals the picture can then be relayed by more conventional communications satellites to the USAF satellite-control center in Sunnyvale, Calif.

By this means, many of the advantages of human judgment and intuition, originally expected to derive from human astronauts in the Manned Orbiting Laboratory, could be obtained with earth-based observers comfortably stationed in terrestrial control centers. It is not known for certain whether Big Bird will have this "real-time" reconnaissance capability. If it does not, such real-time control is almost certain to be included in the fifth generation of unmanned satellites, currently identified by the code number 1010.* The present timetable calls for the USAF to request industry proposals for the 1010 spacecraft during early 1971.

---

* These code numbers are periodically changed for "security" reasons and thus a different number may be assigned by the time this is published.

SATELLITES FOR EARLY WARNING

An important new military mission for satellites had emerged in early 1958 during the period when there were growing American fears of a surprise Soviet ICBM attack, which could wipe out the entire SAC bomber fleet and leave the nation defenseless. The U.S. already had started to build two giant BMEWS radars in Alaska and Greenland, at a cost of $800 million, to obtain a scant 15-minute warning of any Russian missile attack. The precious 15 minutes would enable a small percentage of the SAC bombers, maintained on continuous ground alert with their crews living in nearby trailers, to become airborne before Russian warheads could impact. This would guarantee that the USSR would suffer at least some thermonuclear retribution and thus, hopefully, might deter any thought of such an attack.

If the 15-minute warning time could be stretched to 25 or 30 minutes, many more SAC bombers could become airborne and this would greatly increase the deterrent effect. But the detection range of the BMEWS radars could not be extended further because the curvature of the earth shielded Russian missiles from view until 10 to 15 minutes after they had been fired. With the advent of the Space Age in the fall of 1957, it became possible to consider orbiting satellites over the USSR to detect Soviet missiles almost as soon as they were launched, which could almost double the early-warning time. Several satellites would be needed, suitably spaced in orbit, so that at least one would always be in position over the Soviet Union to maintain a continuous watch. The satellites would use the same type of polar orbit employed for reconnaissance satellites.

By the summer of 1958, Lockheed was under contract to adapt the Agena spacecraft for the new early-warning mission. The program was given the code name of Project Midas, an acronym derived from MIssile Detection And Surveillance. The satellite would be equipped with instruments that could sense the "thermal" infrared radiation emitted by the ex-tremely hot plume of a rocket. To test the concept, infrared sensors were installed in a U-2 aircraft and flown at high altitude over Cape Canaveral during missile launches. As predicted, the infrared sensors seemed able to detect the hot rocket plume.

Spurred by SAC's anxieties about protecting its bomber fleet, the first experimental Midas satellite, with infrared sen-sors built into the nose of an Agena spacecraft, was developed hurriedly. On February 26, 1960, Midas-1 was poised atop an Atlas/Agena at Canaveral, ready for launch, and it was planned to test the satellite against American missiles that would be launched once the spacecraft was in orbit. The launch seemed successful, but an explosion occurred when the time came for the second-stage Agena to separate from the Atlas. On May 24, Midas-2 was successfully launched into a 300-mile-high orbit. For the first two days its infrared sensors seemed to be operating satisfactorily, but then the satellite communications with the ground fell silent.

At this point, in the spring of 1960, the Discoverer pro-gram had been encountering serious problems due to mal-functions of the Agena components in the harsh, cold vacuum of space. Rather than waste more of the precious Midas pay-loads, it was decided to wait until the basic Agena spacecraft and its attitude-stabilization system had been thoroughly de-bugged in the Discoverer program. By the late fall of 1960, these Agena problems seemed solved and the higher-priority photo-reconnaissance program was progressing well enough to allocate a launch vehicle for Midas-type tests. Discoverer-19 was outfitted with infrared sensors, instead of a recoverable capsule, and on December 20 was launched into a low-altitude near-polar orbit from Vandenberg. Again, on February 18, Discoverer-21 was orbited for similar Midas tests.

For more than a year the USAF had been pushing to obtain Pentagon approval to shift the Midas program from its developmental status into a higher-priority production and deployment category. But in the spring of 1961 the new administration's Defense Secretary, Robert S. McNamara, publicly expressed doubts over the feasibility of the Midas concept during Congressional hearings. "There are a number of highly technical, highly complex problems associated with this system," McNamara said. "The problems have not been solved, and we are not prepared to state when, if ever, it will be operational."

The basic problem, beyond unreliability troubles that then plagued all satellites, was that the infrared sensors could mistake the infrared radiation from sunlight reflecting off high-altitude clouds for rocket-engine plumes. This meant that a Midas satellite passing over the USSR might mistake a cluster of high-altitude clouds basking in the sunlight for a mass ICBM attack and flash a false alarm back to the U.S.

Even as McNamara was testifying, the USAF was readying two full-fledged Midas satellites for launch and much would be riding on their success or failure. The Midas payload weighed roughly 2,000 pounds, including delicate infrared sensors and complex electronics, and was mounted in the long nose section attached to the Agena. A powerful Atlas first stage was required to launch the Midas into the 2,000-mile near-polar orbit that would be needed for operational use over the USSR to give the spacecraft sensors a wide-spanning view. On July 12, 1961, Midas-3 was successfully launched into orbit, with an apogee/perigee altitude of roughly 2,100 miles and an inclination of 91 degrees, from Vandenberg AFB, Calif.

The USAF disclosed that Midas-3, as well as Midas-4

which went into a similar orbit on October 21, would be tested against missiles fired from Cape Canaveral and Vandenberg, as well as against special flares designed to mimic the infrared characteristics ("signature") of rocket engines. It was shortly after the Midas-4 launch that the Kennedy administration dropped the heavy security cloak over the reconnaissance satellites, and it enveloped the Midas program as well. But from informed observers it was learned that the Midas was still encountering the same problem of positive identification of missiles and false alarms. It was clear that much more experimental data, and testing, were needed to devise sensors which could discriminate rocket-engine plumes from sunlight bouncing off clouds.

By the fall of 1961 the once-urgent need for additional warning time of Soviet missile attack had eased somewhat because the U.S. now knew that it was the USSR that was on the short end of the Missile Gap. Furthermore, an increasing portion of America's fast-growing missile arsenal soon would be protected against surprise attack in hardened underground silos or in roving Polaris submarines. Additionally, the Kennedy administration had authorized an airborne alert for one-eighth of the SAC B-52 fleet, assuring that this portion could not be caught on the ground and destroyed by a surprise attack.

And so, in late 1961, the Midas program was cut back to a lower-priority effort, with emphasis on making measurements of infrared signatures from space and devising more foolproof sensors. On April 9, 1962, the USAF launched an "unidentified" satellite into an orbit with essentially the same characteristics as Midas-3 and Midas-4, using an Atlas/Agena. More than a year later, on May 9, 1963, another "unidentified" satellite went into a Midas-type orbit, followed by another on July 18, 1963. There would be no more of this type orbited for three years. Clearly, the problem of devising sensors with sufficient discriminatory power was a much more difficult task than had been envisioned several years earlier.

A clue to the Midas troubles, and to a shift in technical approach, appeared in Congressional testimony made public in the summer of 1963. Although the censor carefully deleted the name, there is no doubt that it was the Midas program being discussed' by Dr. Harold Brown, then Director of Defense Research and Engineering, and Congressman George H. Mahon, chairman of the House Appropriations Committee and its Defense Subcommittee. (During the discussion, Mahon said the lack of success in the costly program reminded him of the quotation: "He, Midas-like, turned all to gold." Mahon added: "It appears to me that we have been using a lot of gold.")

Brown explained why McNamara had cut back the USAF's request for $190 million for the program to only $35 million: "The way the program was going, it would never produce a reliable, dependable system." Brown said that a total of $423 million had been spent so far on "premature system-oriented hardware" and costly launch vehicles. The reason for the lack of success, Brown explained, was that "the requirements were set too stringent considering how much we knew. We didn't know enough to be able to go for a system as complicated and demanding as the requirements which were set."

With the more modest funding which McNamara had authorized, Brown said, "within a year or two, reorientation of the program may result in obtaining the basic information which will then enable us to go to some other system." This new design approach, he added, *would be a very different system from the one originally proposed.*" (Emphasis added.)

One of the arguments that the USAF had used to muster support for the Manned Orbiting Laboratory was that while infrared sensors might mistake sunlight bouncing off clouds for a Soviet missile attack, human observers on orbital patrol could never make such a misjudgment. Furthermore, top government officials would be much more willing to accept the word of human observers than a signal from an orbiting automaton. This missile-spotting capability of human astronauts was demonstrated during several of the NASA Gemini manned space missions. For example, during the sixty-first orbit of Gemini-5, launched on August 21, 1965, a small rocket-powered tracked-sled was fired on the ground at Holloman AFB, N.M., just as the Gemini passed over. Astronaut Charles Conrad, Jr., reported: "There it goes. We see it. We could see it very well."

It was disclosed that Gemini-5 astronauts carried infrared devices to measure the radiation signatures of rocket boosters, but the purpose of such data (to devise more effective infrared sensors) was not discussed. During the 62nd Gemini orbit, as the spacecraft neared Vandenberg AFB, a Minuteman missile was launched and the astronauts easily spotted it, although they had some trouble tracking it with their infrared instruments. The crew of Gemini-7, orbited December 4, 1965, readily spotted the launch of a Polaris A-3 missile, fired from a submerged submarine. Astronauts Frank Borman and James A. Lovell, Jr., also sighted a Minuteman warhead (reentry vehicle) as it was descending toward its target in the central Pacific and made infrared radiation measurements, as they had done on the Polaris.

If human astronauts could easily spot the launch of a lone missile, certainly they could detect and identify a mass ICBM attack involving dozens or hundreds of weapons. The unanswered questions were how long men in orbit under zero-G conditions could maintain a state of alertness and whether there was any risk of hallucination in this unnatural environment. Additionally, it would be extremely costly to launch enough manned spacecraft to keep the USSR under continuous surveillance and to resupply the craft with fresh crews and consumables.

Sometime during 1965–66, another concept apparently began to emerge for the early-warning mission, especially after experiments showed no easy solution to the problem of foolproof infrared sensors. The idea was to put a television camera into orbit with a telephoto lens capable of providing resolution at least as good as the human eye from 100 miles' altitude. This TV picture could be relayed to human observers sitting comfortably in terrestrial stations. The satellite also could be equipped with infrared sensors to sound the initial alert, after which human observers could study the TV pictures to see whether the alarm was caused by missile attack or innocent targets.

By the mid-1960s a novel type of orbit for satellites had been demonstrated which had ideal characteristics for such an unmanned early-warning spaceborne system, because the satellite could be made to "hover" over one fixed spot on the earth. In reality, the satellite rotates about the earth along the equator at precisely the same rate as the earth spins about its own axis. This unique orbit, called a "synchronous equatorial orbit," requires that the satellite be boosted to an altitude of about 22,300 miles. This extremely high altitude reduces the amount of payload that can be carried, especially since the satellite must execute a "dogleg" maneuver to get into equatorial orbit from launch sites available in the U.S. To illustrate the payload limitations, the latest version of the Atlas/Agena-D can orbit 9,000 pounds into a 100-mile altitude from Cape Canaveral, or 7,200 pounds from Vandenberg into a near-polar orbit. But the same booster can orbit only 2,600 pounds of payload into a synchronous equatorial orbit.

The first attempt to launch a satellite into synchronous equatorial orbit was made on February 14, 1963, when NASA boosted a small communications satellite, called Syncom-1, built by Hughes Aircraft Co., into a 22,300-mile-high orbit. However, an on-board explosion severed all communications with the satellite. On July 26, 1963, Syncom-2 was successfully placed in a synchronous equatorial orbit. (Once a satellite is in such an orbit, it can be caused to drift to any desired longitude and then can be "parked" there.)

At an altitude of 22,300 miles, a satellite can "see" almost half of the earth. But for an early-warning satellite's sensors to have a more direct view of the vast USSR, it is better to use two satellites—one parked near Singapore to view Siberia and Red China and the other near Ethiopia to scan western Russia. There is one disadvantage of a synchronous equatorial orbit for this type of mission which stems from the fact that part of the USSR is situated at high latitudes where the satellite sensors would have a very oblique view.

There is a simple solution to this problem. If the satellite is launched into a slightly inclined orbit—of, say, 10 degrees— at 22,300 miles, instead of along the equator, then the spacecraft will trace out a figure-8 path over the earth instead of hovering in one spot. The satellite track will loop upward to a latitude of 10 degrees north, then turn and head south, cross the equator and go down to 10 degrees south latitude, finally circling back to the north. The figure-8 ground track will be repeated at the same longitude once every 24 hours.

As the satellite moves north of the equator, its sensors gain a more direct view of Russia's high-latitude regions, but during the next 12 hours the sensors will have a worse view while the satellite is south of the equator. To get around this constraint, another satellite can be launched into an identical orbit at the same longitude except that it lags the first by 12 hours in its path. Thus, when one satellite is north of the equator, the other will be south, and the two will alternate positions every 12 hours.

On November 5, 1967, NASA launched an application-technology satellite, ATS-3, built by Hughes Aircraft, into a synchronous equatorial orbit to perform a variety of experiments. Of special interest here was its TV camera, which would return the first color photos of the earth taken from synchronous-orbit altitude. The ATS-3 camera provided a resolution of approximately 2.2 miles, very adequate for NASA's meteorological experiments but not good enough to spot ICBMs in flight. Since the dramatic ATS-3 photos would be made public, perhaps the Pentagon did not want high-resolution pictures published which might provide a clue to its plans for an early-warning satellite system. Much better resolution could be obtained by using a longer focal-length lens on the TV camera.

Nine months later, on August 6, 1968, the USAF, using an Atlas/Agena, launched a new class of "unidentified" satellite from Canaveral into a near-synchronous orbit with a 10-degree inclination—the first time such an orbit had ever been used for military satellites. This orbit would cause the spacecraft to trace out a figure-8 path at a fixed longitude. The secrecy invoked on this satellite was even tighter than that for the recoverable and radio-transmission-type reconnaissance satellites. Whereas reconnaissance-spacecraft orbital characteristics are published by NASA in its periodic Satellite Situation Report, based on data supplied by the North American Air Defense Command (NORAD), absolutely no orbital data was published for the "unidentified" satellite launched on August 6. In its place the report stated: "Current Elements

Not Maintained." This phrase normally is used for satellites that have ceased to function or those that have exploded in orbit and produced debris that is not worth tracking. As applied to the new type of secret satellite, the statement was not true.

The U.S. government kept its earlier commitment and reported the orbital data to the United Nations, where it is available to the Soviet Union. But the data would have been denied the American public if it were not for Britain's Royal Aircraft Establishment's monthly satellite summary which provided orbital parameters.

This "Advanced Midas" probably was stationed over the Pacific, at least initially, so it could be tested against missiles launched from Vandenberg AFB. A second satellite of this type was orbited on April 12, 1969, also with an Atlas/Agena-D from Cape Canaveral (now called the Eastern Test Range), whose location is better suited for a near-equatorial orbit than the Western Test Range. The second satellite's orbital parameters were nearly identical with the first one's, with apogee of 24,500 miles and perigee of 20,400 miles.

A few cryptic references to the new type of satellites appeared in censored Congressional testimony by Dr. John S. Foster, Jr., the Pentagon's chief scientist, released in late 1969: "We are accelerating the [deleted] program which will provide early warning by satellites . . ." The testimony disclosed that the Pentagon was seeking $157 million for fiscal 1970 for "continued development and initial deployment" of the new Satellite Early Warning System. (The following year, the Pentagon would request approximately $200 million.)

On June 19, 1970, an early warning satellite was launched but failed to reach the required altitude. On August 31 another was fired into an orbit with the usual parameters. Both were orbited by an Atlas/Agena-D plus an unidentified third stage, indicating a heavier, more advanced spacecraft. British data does not show where these two satellites are "parked," but a logical location would be over the western edge of the Indian Ocean to evaluate their performance against rockets launched from Tyuratam and Kapustin Yar.

An improved model of the early-warning satellite, identified by the code number 647, was launched on November 6 using a more powerful booster, the Titan-3C/Agena-D. The satellite, according to RAE data, went into a more elliptical orbit than usual, with an apogee of 22,400 miles, a perigee of 16,200 miles and a period of 20 hours. Whether this resulted from a malfunction or is a more useful orbit for early warning will be revealed as the U.S. launches new satellites to provide continuous coverage over both western Russia and Red China/Siberia, as well as over the Atlantic and Pacific oceans to detect submarine-launched missiles.

The current-design early-warning satellite, built by TRW Inc. (formerly Thompson Ramo Wooldridge Inc.), probably weighs close to 2,500 pounds. The spacecraft's infrared sensors, designed to detect the telltale infrared radiation from ICBM rocket plumes which extend for a mile or more in the near vacuum of space, are supplied by Aerojet-General. The TV camera with telephoto lens reportedly is built by RCA.

If the infrared sensors detect what seems to be the characteristic infrared radiation from ICBMs, an alarm is radioed to a recently completed ground station near Alice Springs in central Australia and to a station on Guam. Either station can relay the alarm instantly, via communications satellites, to Washington, to NORAD Headquarters in Colorado and to SAC Headquarters near Omaha.

If an alarm is received from the infrared sensors, then the satellite reportedly can be commanded to start transmitting a continuous TV picture so that ground observers can determine whether the infrared radiation comes from ICBMs or spurious sources. In the absence of any alarm from the infrared sensors, the satellite probably is designed to transmit a TV picture at periodic intervals, perhaps once every 30 seconds, to check on the satellite's performance and monitor the situation visually. From the spacecraft's altitude of 22,300 miles, the resolution of the TV camera would *not* be sufficient to spot military targets on the ground, even those as large as

* On April 2, 1971, Dr. Foster reported "very encouraging progress in the development of an early warning satellite . . ." during the Pentagon official's testimony before the Senate Committee on Aeronautical and Space Sciences.

an ICBM. But resolution would be sufficient for the intended early-warning surveillance function.

The Russians have often publicly expressed fears that the U.S. plans a surprise attack. It therefore would be quite logical for the Soviets to think of developing a similar early-warning-satellite capability, especially after learning of the Midas program in the late 1950s. There is reason to suspect they may have beaten the U.S. to the punch, using a spacecraft which also serves as a communications satellite. The Russian communications satellites, called Molniya, do not employ a synchronous equatorial orbit as do most American communications satellites. The explanation is that the northern portions of the USSR would not be visible to a satellite located along the equator.

For this reason, the Molniya satellites are launched into a highly elliptical orbit, with an apogee of approximately 25,000 miles and a perigee of only 300 miles with a 65-degree inclination. (See Plate 13-B.) The very high apogee occurs over the northern hemisphere and the satellite's 12-hour period means that it makes two complete revolutions around the earth each day. When the very high apogee occurs over the USSR, each satellite remains within view of Russian ground stations for approximately eight hours before the spacecraft whips down to its low perigee in the southern hemisphere. By placing three (or more) Molniya satellites in this type of orbit and spacing them suitably, at least one will always be within view and available for use.

The satellites have a second apogee each day, and *this occurs over North America*. If the satellites have been spaced to provide continuous communications coverage during the earlier apogee over the USSR, then one of the Molniya satellites will always be in a position to look down on North America. And if these Soviet satellites carried a TV camera, they could be used to warn of any U.S. missile attack.

The Russians launched the first Molniya into this novel orbit on April 23, 1965, and a second into similar orbit on the following October 14. Some weeks after the launch of the third Molniya on April 25, 1966, the Russians casually announced that *the satellite was equipped with a television camera.* The announced purpose was to transmit back cloud-cover pictures.* If a Molniya were transmitting cloud-cover pictures from over North America at a time that the U.S. decided to launch a missile attack, this could instantly be displayed on the consoles in Soviet air defense headquarters and, perhaps, in the Kremlin itself. Since the U.S. has no such intention, the American government welcomes, or at least does not oppose, this Soviet early-warning capability.

If both the U.S. and the USSR have an effective system for detecting a surprise ICBM attack by the other, this is a powerful deterrent against either country's even contemplating such an attack. The reason is that early warning offers the intended victim the possible option, whether or not it is utilized, of launching his entire arsenal of strategic weapons against a preemptive attacker.

Although both the U.S. and the USSR initially gave higher priority to the development of photo-reconnaissance satellites than to the early-warning function, it seems to this author that the latter may prove to have an even greater stabilizing effect on the relations of thermonuclear powers.

---

* I am indebted to Dr. Charles S. Sheldon II, chief of the Library of Congress Science Policy Research Division, for calling my attention to this possible early-warning capability of the Soviet Molniya communications satellites. Dr. Sheldon is one of America's foremost monitors of Soviet space activities outside U.S. intelligence circles. He is one of the few knowledgeable information sources available to the U.S. press, and thus to the American public, on Soviet space capabilities and achievements.

19

NUCLEAR DETECTION AND FERRET SATELLITES

Two other types of satellites have important surveillance missions to perform. One, called Vela (from the Spanish for "watchman"), is used to police the treaty banning nuclear explosions in the atmosphere and in outer space. It also serves to monitor atmospheric tests conducted by Red China and France, which are not parties to the test ban treaty. The other type of satellite, termed a "ferret," is for electromagnetic reconnaissance, primarily to determine the location and characteristics of Communist radars. This type still bears the "top secret" label.

In the fall of 1959 the Pentagon initiated a program to develop new techniques which could be used to police a nuclear test ban, if agreement were later reached with the USSR. Previously, aircraft, ships and terrestrial sensors had been used to detect nuclear explosions in the atmosphere. But with the advent of the Space Age and powerful Soviet rockets, it might be possible for the Russians to conduct clandestine tests far out in space or behind the moon to circumvent the treaty.

It was proposed that American satellites in very-high-altitude orbit might be able to detect such violations. After initial studies showed promise, a contract was awarded to

TRW Inc. in December 1961 to build experimental spacecraft to test the idea. Instrumentation needed aboard the satellite to detect X-rays, neutrons and gamma rays released by a nuclear explosion would be developed by the Atomic Energy Commission's Los Alamos Laboratory and Western Electric's Sandia Corp. One basic instrumentation problem was to be able to distinguish radiation produced by nuclear explosions from similar effects due to natural causes, such as cosmic rays and solar flares.

On October 16, 1963, less than three months after the nuclear test ban treaty was signed in Moscow, the first pair of Vela satellites was launched into a 60,000-mile-high near-circular orbit. The two 300-pound craft were spaced 180 degrees apart around the earth so that they could peer into space from opposite sides simultaneously. The performance of the first pair of Vela satellites exceeded expectations, and a second pair was launched as a replacement on July 17, 1964, into a similar orbit. A third pair was orbited on July 20, 1965.

Although the Soviets have never exploded a nuclear weapon in space in violation of the treaty, so far as is known, the sensitivity of the Vela sensors to cosmic rays and solar-flare radiation has made it possible to tell that the satellites were performing their intended mission. (The Vela data also has scientific value since natural radiation can black out long-range radio communications, can be dangerous to human astronauts and may play a role in the earth's weather.)

In March 1965, a few months after Red China exploded its first atomic bomb, TRW Inc. received another contract to develop an advanced Vela satellite which would be able to measure radiation from nuclear explosions in the earth's atmosphere as well as in space. This would enable the satellite to detect Red Chinese nuclear tests and to measure their yields and other characteristics—beyond performing the original Vela mission. The first pair of the new-design Vela satellites was launched on April 28, 1967, into a near-circular orbit at approximately 70,000 miles. Each of the satellites weighed approximately 500 pounds and carried 138 pounds of instrumentation, 40 percent more than in the previous design.

Defense Secretary McNamara praised the new Vela design during Congressional testimony on February 5, 1968. Referring to Vela-7, -8, launched the previous year, he said: "They have already improved tremendously our detection and yield measurement capabilities."

A fifth pair of Vela satellites was launched on May 23, 1969, and still another on April 8, 1970, each weighing 571 pounds and carrying 163 pounds of instrumentation. A new addition was a sensor to detect the powerful electromagnetic pulse (EMP) generated briefly by a nuclear explosion. These intense pulses can disrupt electronic circuits in missile-guidance systems and other weapons unless protective measures are taken. The Pentagon statement in April 1970 announcing the launch of Vela-11 and Vela-12 said this would "be the last in the Vela series." There was no explanation for the seeming end of this important satellite mission.

The U.S. had not lost interest in policing the test ban treaty or in monitoring Red Chinese nuclear tests. The Vela-type nuclear sensors henceforth would be carried by the new satellite early warning system spacecraft, continuously on station over the USSR and Red China, as described in the previous chapter. This is in keeping with the Pentagon's philosophy of launching fewer but larger spacecraft and using them for multiple functions. The early-warning satellites, it will be recalled, are built by TRW, which also made the Vela spacecraft.

Because the early-warning satellites are cloaked in official secrecy, the Pentagon announcement dared not mention that they would perform the Vela function in the future. The government seemingly fears Soviet protest if it publicly admits that the early-warning satellites are permanently stationed over the USSR even though the Russians are well aware of their existence and purpose, if only from cryptic statements that appear in censored Congressional testimony. However, when the Pentagon's Dr. John S. Foster, Jr., discussed the early-warning satellites guardedly during Senate hearings in March 1970, the censored testimony later released quoted him as saying that they would "take over the test ban monitoring mission currently performed by the Vela satellites."*

---

* A few pages earlier in the released testimony, Foster's comments on the early-warning satellites suffered many more deletions, apparently

The "electromagnetic-reconnaissance" or ferret satellite is also cloaked in secrecy even though the Russians almost certainly operate the same type of spacecraft under the Cosmos label. The mission of the ferret satellite is similar to that performed by the ill-fated USS *Pueblo*—captured on January 23, 1968, with much fanfare by the North Koreans—and to the function performed by numerous Russian "trawler" vessels.

The primary function of the ferret satellite is to pinpoint the location of air-defense and missile-defense radars and to determine their signal characteristics and detection range. This type of intelligence is most important in helping strategic bombers penetrate enemy air defenses, but it also is of value in the design of ICBM warheads to enable them to penetrate ABM defenses. In planning the routes that will be taken by strategic bombers in the event of war, it is vital to know where the enemy's radars are located, their coverage and maximum range of detection. With this knowledge, the bomber routes can be selected to delay the first moment of detection.

Bombers also carry electronic countermeasures equipment, known as ECM, which is intended to interfere with ("jam") or fool enemy air-defense radars. For example, the bomber's ECM may generate spurious signals which create the illusion on the ground radar operator's display that each aircraft is several bombers, each at a different distance. This makes it difficult for the radar operator, and for anti-aircraft missiles, to tell which is the real bomber. The bombers may also carry small decoy missiles, each outfitted with an ECM transmitter which causes the decoy to appear as big as a bomber to the ground radar operator. The decoy missiles can be launched on trajectories quite different from the bomber's intended flight path to divert defending aircraft and missiles.

Before such ECM equipment can be designed and built, however, the characteristics of each basic type of enemy ground radar must be measured. For example, the radar's

---

at the hands of a different censor: "Development of a space-based early warning system is on schedule [deleted]. When operational [deleted] the system [deleted] providing missile attack information [deleted]. Such attack information as [deleted] will be provided . . ."

operating frequency must be determined—the speed at which its antenna rotates, the rate at which it transmits successive pulses and how long each pulse lasts. Naturally, designers of air-defense radars try to devise electronic counter-counter-measures (ECCM) to outwit the bomber's ECM. For example, the ground radar may be designed so that it can quickly change frequency, or pulse repetition rate, to try to unmask spurious targets. This prompts the ECM equipment designer, in his effort to foil this ECCM, to try to endow the bomber's ECM with the ability to instantly change its operating mode to match that of the ground radar.

Until the advent of the Space Age, ferret reconnaissance of air-defense radars was performed principally by aircraft outfitted with elaborate, sensitive receivers. Radar signals were stored on magnetic tape and returned to ground for analysis. To provide the earliest possible warning of a bomber attack, some air-defense radars are installed around the periphery of a country, so the radar signals penetrate several hundred miles beyond the border. This means that ferret aircraft could measure the characteristics of these early-warning radars without penetrating the nation's borders.

But in a very large country, such as the USSR, there are additional groups of radars in the interior which would have to be penetrated in the event of war. Because radars operate at higher frequencies whose range is limited by the earth's curvature (like commercial TV station signals), it is not possible to measure the characteristics of radars deep inside Russia from outside the Soviet Union's borders, even with very high-altitude ferret aircraft. If air-defense radars deep within the Soviet Union have characteristics different from those along its borders, as might be expected, then ECM equipment designed for the characteristics of the peripheral early-warning radars would be useless against those in the interior. One of the important functions of the U-2 flights over the USSR in the 1950s, beyond taking photos, was to perform this type of electromagnetic reconnaissannce.

Another use of ferret reconnaissance is to locate military radio stations, such as those used to direct interceptors to their targets. In the event of war, these would be marked for early attack. Also, eavesdropping on peacetime military communications may provide useful intelligence information. As with radar, such radio communications usually operate at higher frequencies which are limited to line-of-sight ranges.

With the advent of satellites, national borders no longer posed an obstacle to ferret reconnaissance. At an altitude of 300 miles, a satellite would be within line-of-sight range of radar and radio stations up to 1,200 miles away. During one day in orbit, a single ferret satellite would pass within receiving range of all radars within the vast reaches of the USSR and Red China. Accessibility was no longer a problem. But there were many other difficult technical challenges. Aircraft used for this type of ferret work normally carry several human operators whose judgment is needed to operate the complex receiving equipment. Furthermore, aircraft can carry many hundreds of pounds of ferret receivers. In a satellite, space and weight were at a great premium in the early 1960s, and the function of human operators would need to be automated.

The optimum altitude for a ferret-type satellite is slightly higher than for a photo-reconnaissance spacecraft, for which the lowest possible altitude is desired to provide maximum possible ground resolution. The ferret type has no film to be consumed, so it can have a relatively long useful life if its altitude is above 200 miles. Then the practical limit on useful life is determined by the life of its batteries and solar cells, and the reliability of its complex electronic receivers and the tape recorders which store the received signals. Later, when the ferret satellite passes over a friendly ground or shipboard station, the recorded signals are transmitted to earth.

However, a ferret satellite ought not be placed in too high an orbit because this unnecessarily decreases its sensitivity. (The strength of the received radio/radar signal decreases with the square, or second-power, of distance between the signal source and the satellite.) Furthermore, the higher the satellite orbit, the less payload that can be orbited by any launch vehicle. These factors suggest that an optimum altitude for most ferret missions would be in the range of 200 to 300 miles, providing an orbital life of several years.

It is not as easy to positively identify the ferret satellites. Unlike the Samos, whose early launches were identified as such, or where the spacecraft remains in orbit only briefly to mark it as a recoverable type, there were no launches identified as ferret types prior to the invoking of the "top secret" security in the fall of 1961. Nor was there any published discussion of ferret-satellite plans in Congressional hearings, as for the photo-reconnaissance type. Historically, ferret-type missions always have been more sensitive than aerial reconnaissance.

In all probability, some early ferret-type experiments were conducted in the Discoverer series and perhaps as a "hitchhiker" aboard some of the radio-transmission-type spacecraft. The first satellite whose orbital parameters suggest it may have been a full-fledged ferret spacecraft is one launched on May 15, 1962, using a Thor/Agena-B. The satellite went into a near-polar orbit with a perigee of 180 miles and an apogee of 401 miles. On June 18 an "unidentified" satellite was launched into a slightly higher, more circular orbit having characteristics which seem better suited for the ferret mission. The perigee was 234 miles, the apogee was 244 miles and the inclination was 82 degrees. The launch vehicle was also a Thor/Agena-B.

Intuition suggests that a family of complementary types of ferret satellites could be used effectively, much in the pattern of the photo-reconnaissance spacecraft. For example, one type would be used for large-area surveillance, to locate and log the approximate positions of Soviet radars and to determine the frequency band in which each operates. A second type of ferret satellite, larger and more complex, could then be used to make a more detailed analysis of the characteristics of each basic-type radar in the Soviet inventory. Analysis of the ferret-satellite launches starting in 1963 appears to confirm this deduction.

On January 16, 1963, what is presumed to be one of the smaller-type area-surveillance ferret satellites was launched, using a Thor/Agena-D, into a nearly 300-mile circular orbit. It would stay in orbit for nearly six years, but almost certainly would not remain operable for this long period. On June 29 it was joined by a heavier, more complex spacecraft, also in a 300-mile near-circular orbit, which had been launched by the more powerful TAT/Agena-B. It is clear that the satellite was considerably heavier and more complex than previous ones launched with an ordinary Thor/Agena, because the added cost of the strap-on rockets could be justified only if the extra payload capacity were needed.

The ferret receivers carried in these satellites were built by Airborne Instruments Laboratory, Mineola, L.I., N.Y., a company with considerable previous experience in ferret and ECM technology. (The company now is the AIL Division of Cutler-Hammer Corp.)

A curious pair of "unidentifieds" was launched into a roughly 200-mile orbit on August 29, using an ordinary Thor/Agena, and was followed on October 29 by a similar tandem launch using the more powerful TAT/Agena. Again, on December 21, another pair of satellites was launched from Vandenberg, but this time only one of the spacecraft went into a 200-mile orbit while the other went into a lower orbit with a perigee of only 107 miles—in the "photo-reconnaissance region." Could this have been a malfunction involving premature separation of one of the satellites, or was it a recoverable photo type which had been paired with a ferret simply for economy of launch? It was impossible for an "outsider" to know. One could only examine subsequent launches and look for a possible pattern.

During 1964 the USAF stepped up the launch pace for ferret-type satellites as it sought to catalogue the Soviet air-defense radars. Five of the larger, more complex spacecraft were launched individually into 200–300-mile orbits in February, June, July, August and November. Then, on July 6, another of the curious "odd-couple" pair of satellites was launched, this time using the Atlas/Agena-D with its larger payload capacity. One of the two appeared to be a ferret type, judged from its altitude, while the other had a perigee of only 75 miles and remained aloft for only two days—suggesting it was a recoverable photo type. Again, on October 23, an Atlas/Agena was used for a multiple-satellite launch to place two satellites in ferret-type orbits while a third went into an 86-mile perigee orbit, characteristic of a photo mission.

The pieces of the puzzle began to fit together in 1965, when the USAF launched pairs of satellites on April 28, June 25 and August 3, using an Atlas/Agena-D. In each instance, the second-stage Agena went into a low-perigee orbit of less than 100 miles while a smaller satellite apparently was boosted by a self-contained rocket into a nearly circular orbit of approximately 300 miles, characteristic of the ferret mission. On the basis of radar measurements, the Royal Aircraft Establishment estimates that these smaller satellites measure approximately 3 feet in diameter and weigh roughly 125 pounds. This small payload suggests that the satellites were of the type used to make a simple inventory of Soviet radars.

During all of 1965, there appears to have been only a single launch of a ferret satellite of the heavier, more complex design. This occurred on July 16, 1965, using a TAT/Agena-D, with the satellite going into a 300-mile circular orbit. The paucity of launches suggests that spacecraft payloads were performing with increased reliability. (Until 1964–65, tape recorders had exhibited very short orbital lifetimes of around 100–200 hours.)

The pattern continued in 1966. There were three of the tandem launches, with a small ferret satellite seemingly paired with a recoverable photo satellite—on May 14, August 16 and September 16. In addition, there were two launches of the heavier-type ferret, orbited individually by TAT/Agena-D vehicles, on February 9 and on December 28. At the 300-mile altitudes being achieved, these satellites would remain aloft for several years.

In the summer of 1966, it will be recalled, the U.S. introduced the new and more powerful Titan-3B/Agena-D booster to orbit the new generation of heavier, more sophisticated recoverable photo-reconnaissance satellites, previously launched by the Atlas/Agena. The Atlas/Agena would make its last launch of a recoverable photo satellite in June 1967, and so other provisions would have to be provided to launch ferret satellites in the "piggy-back" manner introduced in 1964. (At about the same time, a new generation of ferret receivers, reportedly produced by Sanders Associates of Nashua, N.H., would be introduced.)

The "piggy-back" launch mission had been reassigned to the venerable Thor, whose length had been increased by nearly 20 percent to accommodate more fuel and provide a longer thrusting time. With three strap-on solid rockets, it was called the Long-Tank-Thrust-Augmented-Thor (LTTAT).

The first operational use of the LTTAT/Agena-D had occurred on August 9, 1966, when the new booster was used to
orbit a radio-transmission type of photo-reconnaissance satellite. The next LTTAT/Agena-D launch came on May 9, 1967,
only a month before the Atlas/Agena was retired, and it carried a pair of satellites. One went into the low-perigee orbit
used for recoverable photo satellites while the other went into
a 345/500-mile orbit suitable for the ferret mission.

After this date, many of the LTTAT/Agena launches carried pairs of satellites with one spacecraft going into a very
low-altitude perigee while the other was boosted into a near-
circular orbit at approximately 300 miles. During 1967 there
were three such launches: on May 9, June 16 and November
2. In addition, the USAF launched one of the heavier ferret
satellites on July 24—the only one of this type orbited during
1967, so far as is known. (Presumably, by this time the U.S.
had logged the characteristics of existing Soviet and Red
Chinese radars and now needed only to monitor for new or
improved models.)

A smaller ferret satellite launched in tandem fashion on
December 12, 1968, and another one of this type orbited on
February 5, 1969, went into a much higher than normal altitude—roughly 900 miles. This was about the time that the
anti-ballistic missile (ABM) system which the Russians were
building near Moscow appeared to be reaching an operational status. The higher altitude of these ferret satellites
would be preferable for measuring the coverage of Soviet
ABM radars.

During 1968 there were five of the tandem launches of
photo and smaller ferret satellites, and two of the larger ferret
spacecraft were orbited, one early in 1968 and the other in
the fall of that year.

When development of the Manned Orbiting Laboratory
started in 1965, it was envisioned that the astronauts could
speed up the acquisition of electromagnetic intelligence, as
well as photo reconnaissance, through on-the-spot decision
making. Each radar signal detected could be checked against
a catalogue to see if it was a new site or new type of signal.
If new, the astronauts could select suitable equipment to analyze the signal during the same pass and report the informa-

tion within an hour as the spacecraft passed over friendly shores. With the then-current designs of unmanned ferret satellites, two different types of satellites were required to perform this function, with intervening analysis on the ground.

However, because of uncertainties over the future of the MOL program, even at the time it was started, work was also begun on a third-generation unmanned ferret satellite, much as had been done for a new generation of unmanned photo spacecraft. Hughes Aircraft Co. reportedly was selected as the prime contractor for the satellite payload on the basis of the company's considerable experience in building communications satellites for commercial service.

It seems likely that the new ferret satellite will be able to handle both the initial surveillance and the detailed signal-analysis functions. The spacecraft probably will carry an inventory of all known Soviet radars, their locations and signal characteristics in an electronic computer. If a new site or new type of signal is detected which does not appear in the inventory, the satellite will automatically analyze the characteristics during the same pass and take an electronic bearing to the radar to fix its location.

The first launch of the new heavy ferret satellite is expected in 1971. The choice of launch vehicle, possibly a Titan-3B or even a Titan-3D, will indicate the weight of the new design and provide a clue as to its sophistication and potential capabilities.

20

FIRM NUMBERS REPLACE "INTELLIGENCE ESTIMATES"

The sharp debate in the U.S. over whether or not to deploy an anti-ballistic missile (ABM) has provided a showcase for intelligence obtained by satellites, although the source of the information is never mentioned officially.

Since the mid-1950s, when the U.S. began to develop an ABM called Nike-Zeus, three different administrations had resisted pressures to deploy the system because it was believed that the ABM could be overwhelmed by a massive Russian attack. In the early 1960s, Russian leaders had boasted that the USSR had solved the ABM problem, and models of ABM missiles were publicly displayed in November Day parades in 1963 and 1964. In 1965, Soviet television showed movies of what were said to be ABM intercepts of a target missile. Still the U.S. held back on ABM production.

But during late 1965 and early 1966, American satellite photos showed that the Russians were constructing some type of defensive missile system around Moscow. The key question was whether this was merely an improved defense against SAC bombers, or the first operational Soviet ABM. By the fall of 1966, satellite photos had eliminated all doubt. On November 10, Defense Secretary McNamara, following a meeting with President Johnson, held a press conference to reveal that the U.S. had evidence that Russia was now deploying an ABM system around Moscow.

The U.S. government sought to prevent a costly ABM arms race. President Johnson's State of the Union message in January 1967 proposed joint talks on ABM limitations and he followed this with a personal letter to Soviet Premier Kosygin. In Moscow, American Ambassador Llewellyn E. Thompson actively tried to arouse Russian interest. Meanwhile, McNamara's 1967 Posture Statement sought to alert the Russians to the futility of a costly ABM investment by revealing that the U.S. had developed multiple warheads for its ICBMs and Polaris-type missiles. The multiple warheads would make it still easier for ballistic missiles to saturate and overwhelm an ABM defense.

But the message was slow to penetrate, perhaps because the Russian military have always been oriented toward strong defensive measures. On February 10, 1967, during Premier Kosygin's visit to London, he was asked by the press for his views on an ABM arms race. Kosygin responded with a strong defense of ABMs. During the next eighteen months, the Russians continued to construct ABM sites around Moscow until there were approximately 60 launchers in place. By the fall of 1967, they also had more than doubled their arsenal of ICBMs, with 720 deployed, compared with only 340 a year earlier.

On September 18, 1967, McNamara disclosed that the U.S. had decided to deploy a limited ABM system. He indicated that the decision had been made reluctantly because no ABM could provide a complete defense of large population centers against a massive attack. But he said the situation was "analogous to the one we faced in 1961 . . . we are uncertain of the Soviets' intentions."

When the new Nixon administration took office, it embraced the ABM with much more fervor than its predecessor. As the ABM debate grew more heated, the new administration released more details on the growing Soviet strategic power—details which could come only from satellite photos. During Congressional testimony on July 16, 1969, for example, Defense Secretary Laird revealed the Russians were building a new type of nuclear-powered submarine, designed to carry

16 Polaris-type missiles. "We now know that this submarine (designated the Y-class) is in full-scale production at a very large facility near Archangel–Severodvinsk–and possibly at another smaller yard. These two facilities can accommodate a total of 12 complete hulls . . . Eight or nine Y-class submarines have already been launched and several are believed to be operational," Laird disclosed.

The Pentagon chief also revealed that "the Soviets have just started to deploy a new solid-fueled ICBM, the SS-13." More disquieting, the Russians now had more than 230 of the large SS-9s either in operation or under construction, Laird reported. (The SS-9, which can carry three 5-megaton warheads, would be the weapon used to destroy American Minuteman missiles in their underground silos if the Soviets were to launch a surprise attack.)

During the same hearings, the Pentagon's chief scientist, Dr. John S. Foster, Jr., disclosed that the Red Chinese were building facilities to launch a full-scale ICBM. American satellite photos also showed the extent of Soviet fears of Red China, in the wake of the border clashes that had occurred during 1969. Originally the Russian radars around Moscow and at the Russian ABM test site had been aimed toward the polar routes over which the U.S. missiles would travel. But on June 16, 1969, Laird revealed in Congressional testimony that the Soviets "are now going forward at their test sites and are aiming their radars at China, and they are also changing the configuration of the radars around the Moscow complex. *We have very good evidence—solid, firm, hard intelligence estimates—that they are pointing these radars toward China* [emphasis added]."

The dramatic change that has occurred since the late 1950s was highlighted by an exchange between Dr. Foster and Senator Stuart Symington on June 13, 1969, during Senate Appropriations Committee hearings. It will be recalled that Symington had been one of the most concerned, and outspoken, critics of the Eisenhower administration's defense policies in the late 1950s:

FOSTER: I remember a period in the late 1950s when we were very uncertain about what Russia was doing. We did not have sufficient intelligence, and we were concerned that she might be constructing a large number of ICBMs.

SYMINGTON: We were very concerned and I was caught. The result was that testimony before this Committee projected the number of bombers that they were [expected] to build to the thousands and thousands. Yet they didn't build anything remotely comparable to what was said they would build. That was in the early 1950s . . . In the late 1950s, and again I was caught, we estimated the number of ICBMs they had on launching pads; and *between December 1959 and August 1961, this Government cut the estimated number of Soviet ICBMs on launching pads by 96½ percent*. So both of the apprehensions in the 1950s, and I blame myself as well as others, were almost 100 percent wrong in the estimates of the danger. [Emphasis added.]

FOSTER: I think, Senator, it is very fortunate that today we have far better intelligence, so we argue about relatively small uncertainties in the intelligence, but we are not going through the great spasm of building missiles . . . That is why in equivalent dollars, the amount of money we are requesting for operation and deployment of strategic systems [weapons] is less than half of what it was in the very early 1960s.

In early 1970, as the Nixon administration sought to expand the Safeguard ABM system deployment, thereby sparking more heated controversy, even more specific details obtained from satellite photos were made public to bolster the administration's position. In the summer of 1969, Laird had given a lumped total for the number of SS-9 missiles in operation and under construction. But on April 20, 1970, when the Defense Secretary spoke to the Associated Press editors in New York City, he was more specific. There were, Laird said, 220 operational SS-9s and approximately another 60 sites under construction. (He reminded his audience that five years earlier there were no SS-9 missiles in operational readiness.) He reported that the Russians had more than 800 of their smaller SS-11 and SS-13 ICBMs deployed, whereas five years before there were none.

Laird revealed that the Russians then had 64 ABM missile launchers in place around Moscow, together with associated early-warning and guidance radars. (Several months later, Dr. Foster would disclose that the Russian early-warning radars were as big as "three football fields laid end-to-end," and resembled a giant "hen-house.") The Defense Secretary even catalogued Soviet Polaris-type missile strength in terms of the number of launchers, rather than the number of submarines. He said the Soviets had more than 200 missile launchers aboard nuclear-powered submarines and roughly 70 launchers aboard diesel-powered types.

What Laird did not mention was that satellite photos indicated that the Russians had not started building any new missile sites since the start of the Strategic Arms Limitation Talks (SALT) in Helsinki on November 17, 1969. (The Soviets had finally responded to President Johnson's overtures in June 1968, but the Russian invasion of Czechoslovakia on August 20, 1968, had delayed the start of formal negotiations.)

The Soviet moratorium on new missile-site construction appeared to continue as late as June 24, 1970, when the chief negotiator at the SALT meeting, Gerard C. Smith, returned to Washington for talks with President Nixon. On that date a high government official indicated privately that there were no new Russian missile starts—an encouraging sign. But on July 8, just fourteen days later, Defense Secretary Laird told a press conference that the Russians had started to construct new missile sites, including ones for the large SS-9s. This seemed to contradict official views expressed only a few days before. The explanation was that the U.S. had new satellite photos. A recoverable-type photo satellite had been launched on June 25, using a Titan-3B/Agena. On July 6, two days before Laird's press conference, the film-pack had been returned from orbit, recovered near Hawaii and flown to Washington for an analysis at the National Photographic Interpretation Center.

Within 48 hours after the film had been kicked out of orbit, top U.S. officials knew that the Russian missile-site moratorium had ended, for unknown reasons, and the U.S. public learned this disquieting piece of news the following morning in the press. Within several days the Associated Press obtained more details. The new construction consisted of three "flights" of SS-9s, each consisting of six launchers.*

For those of us who believe the time has come for the secrecy wraps to be *partially* lifted from the reconnaissance-satellite program, this was an encouraging sign. Yet, little more than a month later, the administration would be much too secretive about satellite reconnaissance, in the hope of not upsetting a tenuous cease-fire in the Middle East—only to have its strategy backfire.

Under the terms of the cease-fire which took effect one hour after midnight on August 8, no new weapons were to be introduced into a zone 50 kilometers (31 miles) wide on either side of the Suez Canal zone. Within several days after the truce, Israel charged that new Soviet anti-aircraft missiles had been brought into the zone in violation of the agreement, and made public aerial photos to support its charge. The U.S. at first tried to ignore the charges, but on August 19 the U.S. State Department finally responded with an official statement that "our evidence of this is not conclusive." Unofficially, it was explained to the press that the U.S. had not gotten around to sending a U-2 from nearby bases in Turkey or West Europe to take photos *until two days after* the cease-fire went into effect.

If the U.S. expected minor violations during the first couple of days and wanted to be in a position where "officially" it appeared unable to confirm nor deny such charges, it would be diplomatically convenient to "forget" to dispatch a U-2 until several days after the truce began. But if satellite photos could be secretly obtained prior to and following the cease-fire, top U.S. officials could really know what was taking place without having to admit they knew.

On July 22, a little more than two weeks *before* the cease-fire, the U.S. launched a radio-transmission-type reconnaissance satellite from Vandenberg, but it did not go into the

---

* In mid-December 1970, Laird released a statement that spoke of "some preliminary indications" that the Soviets were slowing down on their construction of new SS-9 missile silos. The evidence came from photos obtained from a recoverable reconnaissance satellite launched on October 23, using a Titan-3B/Agena-D.

customary near-polar orbit. Instead, the LTTAT/Agena-D executed a dogleg maneuver which put the spacecraft into an orbit with an inclination of only 60 degrees—lower than had ever before been used for such a satellite. And this 60-degree-inclined orbit took the satellite directly over Syria, Israel, northern Sinai, the Suez Canal zone and northern Egypt, heading south-by-southwest once each day.

Another curious feature was that the satellite was launched from Vandenberg at approximately 6 P.M. local (standard) time, whereas such spacecraft are launched between 10 A.M. and 2 P.M. for normal missions. This unusual launch time would bring the satellite over the canal zone at around 6 P.M. local time, when the sun would be low and cast long shadows, making it easier to spot small targets such as anti-aircraft batteries. The satellite also makes a second pass each day at around 6 A.M., in a north-by-northwest direction. In the cool early morning hours, it would be easier for the satellite's infrared scanner to spot warm vehicle engines against the cold desert sands if the vehicles had been used during the night to shift missile batteries.

(I discovered the unusual features of the July 22 satellite by chance, during research for my book. While perusing the July 31, 1970, issue of the NASA Satellite Situation Report, my curiosity was piqued by the unusual 60-degree inclination for the reconnaissance satellite—which prompted me to plot its ground track.)

My article disclosing that the U.S. had placed a radio-transmission-type reconnaissance satellite into the peculiar orbit which took it over the Middle East battle zone was published in *Aviation Week & Space Technology* on August 31. Two days later, *The Washington Post* reported that "the United States now has clear evidence that Egypt and the Soviet Union have been cheating on the cease-fire agreement in the Middle East." The newspaper said that President Nixon had been advised of this the day before (September 1). On September 4, during a State Department press briefing on the Middle East, *New York Times* correspondent Tad Szulc, in an attempt to get a subtle confirmation or denial of my satellite article, asked whether the U.S. had operated *any* type of surveillance vehicles over the battle zone *before* the truce. The State Department official declined to comment.

The idea of performing advanced reconnaissance secretly by satellite, while creating the illusion of a 48-hour delay because of belated U-2 missions, must have seemed appealing to those responsible for the idea. The U.S. could not fly a U-2 over the area with complete safety until after the cease-fire. And to have advised the three principals of the reconnaissance-satellite mission would almost guarantee that the fact would be revealed in the Egyptian press, thereby compromising the long-standing "top secret" security classification.

Furthermore, disclosure of the satellite mission might well bring a scathing denunciation from the Egyptian press, perhaps even the Russians, charging that U.S. satellite photos would be given to Israel if fighting should resume. Less than a year earlier, the Prime Minister of Sudan, who is very friendly to Egypt, had lambasted American reconnaissance satellites before the United Nations. On September 24, 1969, Babiker Awadalla told the UN: "Some of us cannot but feel that the world would have been a better place if it were free of American orbiting spies in the sky, free of their intelligence ships, free of . . ." (Awadalla made no mention of Soviet reconnaissance satellites or ferret "trawler" ships.)

From the vantage point of hindsight, if the U.S. had been less secretive about its plans for satellite reconnaissance over the truce area, this in itself might have discouraged any temptation to cheat and thereby have prevented the complications which the violations precipitated. This incident holds important lessons for both American and Soviet policy makers and military strategists, especially in view of the reported progress toward some agreement to limit strategic weapons at the SALT negotiations in Helsinki and Vienna.

Reconnaissance satellites will be the principal means of policing any agreement to limit strategic weapons. If and when such agreements are presented to the U.S. Senate for treaty ratification, and thus to the American public for approval, the "top-secret" cloak of secrecy, which has for ten

years attempted to hide even the existence of the reconnaissance-satellite program, will have to be eased. With the
recent Soviet violations of the Middle East cease-fire, there
will be an unwillingness to accept any arms treaty with
Russia that appears to depend on trust alone.

I do not propose that every detail on all operational capabilities of our latest designs should be made public. But the
existence of such satellites, their general capabilities *and their
inherent limitations* must be made known to the public and
to the Congress. For if satellite photos should ever reveal
Soviet violations of strategic arms-control treaties, the Congress and the American public must face up to the implications of such actions.

By lifting, at least partially, the official veil of secrecy that
has obscured America's reconnaissance and early-warning
surveillance-satellite capabilities, the U.S. government can
help to assure that they are known not merely to top Soviet
officials but throughout that country's vast civil and military
organization, to help discourage any attempt at even minor
violations of arms-control treaties. As events in the Middle
East have demonstrated, it is easier to discourage violations
than to expose them later and try to get the offender to publicly admit wrongdoing and then to take corrective action.

With the existence of both American and Soviet reconnaissance-satellite capabilities out in the open, if violations
do occur there would no longer be any reason for withholding
photographic evidence. Such evidence could be used to
muster world opinion, to which major powers are not insensi-
tive. All parties to a strategic arms control treaty should
know this. And the leaders of smaller nations, including
Sudan's Babiker Awadalla, need to understand that the world
is a safer place *because* of the prying satellite eyes in orbit.

In the fall of 1962, when the U.S. publicly accused Russia
of installing ballistic missiles in Cuba, Soviet diplomats ve-
hemently denied the charge. These denials quickly ended
when President Kennedy decided to make public the reconnaissance photos showing Soviet missiles in Cuba, despite objections by some advisers that this might compromise our intelligence sources. Shortly thereafter, the Russian missiles were
withdrawn. More important, because the Russians know that

we are continuing our aerial reconnaissance, they have never repeated their mistake.

The lesson, as applied to satellite reconnaissance, should be clear. It was succinctly stated centuries ago in a Chinese proverb: One picture is worth ten thousand words.

21

THE INHERENT LIMITATIONS OF
SPACEBORNE RECONNAISSANCE

The dramatic accomplishments of reconnaissance satellites during the first decade of their existence must not obscure their inherent limitations, because this could lead to an unjustified relaxation of national vigilance. So long as large ballistic missiles and bombers remain the primary means of delivering thermonuclear weapons over long distances, reconnaissance satellites should be able to police any strategic arms control agreement that is likely to emerge from current SALT negotiations. The large-scale, time-consuming construction required to build underground missile silos and support facilities, and the air bases needed for strategic bombers, almost guarantees that any gross violation can be detected by satellites. Construction of missile-launching submarines is only a little less certain.

It is conceivable that a country might try to construct underground missile silos by working entirely beneath the surface so as to avoid satellite detection, but this would be extremely expensive. There is an easier, more cost-effective alternative which the U.S. and the USSR already have developed for other reasons. This is to miniaturize an H-bomb so that each missile can carry several smaller warheads, called

Multiple Re-entry Vehicles(MRVs), instead of a single large weapon. In the simplest design, several warheads are released in shotgun fashion during the ballistic trajectory to increase the chances that one will impact on the target. A more sophisticated version, called Multiple Independently targeted Re-entry Vehicle (MIRV), enables each of the smaller warheads to be directed to a different target in the same general area. The U.S. is beginning to deploy MIRVs on its Minuteman-3 missiles, and the Russians have tested a three-warhead payload for their SS-9s.

Once a missile is deployed inside its silo or submarine, it is impossible for a reconnaissance-satellite photo to show whether it carries MRVs or MIRVs, or how many individual warheads are aboard the missile. However, from the known size of the missile and the thrust of its rocket engines, and the current state of the art in H-bomb miniaturization, it is possible to estimate roughly how many of the smaller warheads can be carried. For example, current U.S. projections estimate that the Soviet SS-9 can carry a single 25-megaton warhead or three 5-megaton MRVs. (The total megatonnage for MRVs always is somewhat less than the figure for a single warhead because each warhead must carry its own detonating mechanisms.)

In time, the U.S. will divulge the number and size of its MRVs and MIRVs, either in Congressional testimony or in its press. The Soviet Union is much less accommodating, but Russian missile tests in the Pacific have revealed that the SS-9 is being readied to carry three MRVs or MIRVs. The latest version of the Polaris, the A-3, carries a three-warhead MRV cluster, and the new Poseidon which will replace it is expected to carry as many as 10 MRVs. It is conceivable that very-high-resolution satellite photos of Russian missiles being transported to their silos may permit determination of whether they are outfitted with multiple warheads. This is one reason why the U.S. has poured so much money into the development at Perkin-Elmer of an ultra-high-resolution camera for the fourth-generation reconnaissance satellite, which will soon make its debut.

The miniaturized warheads being developed for MRVs and MIRVs have broader and more ominous implications. As

H-bomb size is reduced, the size of the missile needed to launch a single warhead also decreases. This, coupled with advances in more powerful rocket fuels, opens the way to relatively small mobile ICBMs which can be housed in large trucks. These could be disguised, for example, to resemble trailer-trucks used to transport gasoline. Thus, the U.S. may soon need to catalogue the number and types of large vehicles within the USSR which may conceivably be used as launchers for small ICBMs. Following any SALT treaty, photo-interpreters will need to be alert to any upsurge in numbers of large vehicles, especially in geographic areas suitable for ICBM launching. This is another mission for the ultra-high-resolution camera destined for the fourth-generation satellites.

Equally important, the U.S. itself must explore new rocket-fuel and weapon technology so that it can estimate how small an ICBM is feasible and how small a mobile launcher is possible. *Without firsthand knowledge of what is technically possible, the best satellite photos may provide scant information of value.*

For example, during the early 1940s, British intelligence received a number of reports that the Germans were developing several types of guided missiles to bombard England, and that one of these was a ballistic missile. But Prime Minister Churchill's personal scientific adviser, Dr. F. A. Lindemann (Lord Cherwell), ridiculed the idea that it was feasible to build an effective long-range ballistic missile. Thus, when British photo-interpreters studied aerial pictures of the German missile-test station at Peenemünde in May 1943, they had no technical background in what to look for, and they found no evidence of missile activities, despite the fact that there were launchers in place for both the V-1 flying bomb and the V-2 ballistic missile. A V-2 missile which appeared on an aerial photo taken on May 14, 1943, was identified merely as an "object." Similarly, the function of large structures that resembled a ski jump, which the Germans were building in northern France to launch V-1 missiles, could not be fathomed by British analysts as late as June 1943.

Yet, within twelve months of this date, the first of hundreds of V-1s would begin to impact on Britain, and three months later the first V-2s would begin to fly their trajectories of terror.

The problem of interpretation persists today, despite the technical advances in aerial/spaceborne photography and interpretation. For instance, the U.S. intelligence community debated internally for several years as to the function of the large radar and defensive-missile network that the Russians were building near their northern border. Some believed it was intended as a defense against manned bombers, while others were sure it was an ABM system. (The network is called the Tallinn system because it begins near the Estonian capital city of the same name.)

As recently as the summer of 1969, the Pentagon's Dr. Foster admitted that the primary function of the Tallinn system was still in doubt. Responding to a question during Congressional hearings, Dr. Foster said: "The evidence available on the Tallinn system is both incomplete and ambiguous. Some of the evidence suggests a SAM [Surface-Air Missile system for bomber defense], while other less comprehensive evidence suggests ABM." By the spring of 1970 the emerging consensus in the intelligence community was that the Tallinn system is intended primarily for bomber defense, but some still believe that it may have a limited ABM capability.

It is important to emphasize an inherent limitation of satellite reconnaissance which seems obvious but whose implications generally are not fully appreciated by the layman. Satellite photos cannot reveal what new weapons may be under development in the laboratory. The earliest possible moment that such developments usually can be discovered by satellite photo is the time when a new weapon is taken outdoors for tests—*if tests outdoors are indeed required.* Even then, the Soviets can easily predict when one of the American reconnaissance satellites will pass within view of the test range and may be able to schedule the tests to avoid detection, unless they involve very long-range missile firings where the re-entry vehicle must impact in the Pacific ocean.

If such precautions are taken, then the earliest possible moment of discovery of a new type of weapon could be delayed until it is in production and ready for deployment at a remote location—*if the weapon needs to be transported to its operational site and cannot effectively be camouflaged during transport.* Even then, the nature and purpose of the new weapon may be difficult to determine from satellite photos, as previously indicated.*

The foregoing is not simply alarmist hypothesizing, for there is a small cloud on the horizon which *could* have a destabilizing influence on Great Power relations. It is a remarkable recent scientific discovery which may develop into a new type of weapon that conceivably could be deployed without being detected by reconnaissance satellites. The device is the "laser," which generates an intense, finely focused beam of unusual light, referred to as "coherent," which even nature itself does not produce, so far as is now known. The laser was theorized in the late 1950s, and the first working model, which used a ruby crystal to generate the unusual light, was built by Dr. Ted Maiman in 1960. (Maiman was a scientist with Hughes Aircraft Co. at the time.)

The development of the laser quickly triggered speculative stories that it might develop into a "death ray"—popularized in the Buck Rogers comic strip in the 1930s. The concept seemed to gain credence from numerous demonstrations which showed that a laser beam could instantly burn a hole through a thin razor blade. This was an impressive demonstration, but the power levels then obtainable were infinitesimal compared with those needed for any practical weapon application. It was as if someone, after witnessing the first flight of the Wright Brothers at Kitty Hawk in 1903, had suggested that man might soon fly to the moon. Yet if anyone had been bold enough to make such a prediction at the time, he might well have lived to see it come true.

Experience shows that technological processes and devices

---

* On March 4, 1971, intelligence officials testifying before the Senate Armed Services Committee disclosed that the Russians had started to build a new type of underground silo but admitted that it was not known whether this presaged the introduction of a new-type ICBM. Presumably this new intelligence had come from a recoverable satellite launched on Jan. 21 whose capsule had been returned on Feb. 9.

can be scaled up in size and power in varying degrees before each reaches a practical limit. Today's jet engine is roughly fifty times more powerful than the first models, while the largest rocket engine today is a hundred thousand times more powerful than some of Professor Robert H. Goddard's earliest models. Neither technology has yet reached its upper limit.

If the laser were ever to become a "death ray," the output energy levels obtained with early models would have to be increased at least a millionfold, perhaps a billionfold. In the very early 1960s the only known way of producing laser action was to use glasslike materials. Because of relatively low efficiency, a vast amount of heat is generated internally to produce even modest output power levels, and so the heat generated at high power levels could destroy the glasslike material. For this reason, as well as other technical obstacles, it seemed doubtful that the laser could ever become a "death ray"-type weapon.

The Pentagon, spurred by scientific papers indicating that the Soviet Union was actively exploring laser technology, funded modest research and exploratory developments with expenditures of about a couple of million dollars a year. Out of this, and industry-funded research, has come a variety of very useful if less glamorous applications for the laser in defense, industry, science and communications. (For example, within several decades, long-distance telephone communications may travel on laser light beams instead of using wires and microwave radio-links. The laser also is being used for retinal surgery and may find other medical uses. Military applications include the use of the laser as an optical radar to measure range, to illuminate the ground for aerial photography and for missile guidance.)

Within several years after Maiman demonstrated the first laser using a ruby, a variety of other materials were discovered that could produce laser action, including gases, such as neon and carbon dioxide. Carbon dioxide not only can generate much higher power levels, but being a gas it can be circulated through cooling mechanisms to dissipate heat generated by the lasing action. More recently, high-energy lasers that utilize chemical reactions have been discovered.

There is evidence that the Soviets are well aware of the potentialities of lasers as weapons. In 1968 two Russian scientists from the Lebedev Physics Institute, N. N. Sobolev and V. V. Sokovikov, published a paper on carbon dioxide lasers which included the following comment: "Finally, different military applications are possible. In particular, targets can be damaged with the aid of a laser beam if powers on the order of several kilowatts (thousands of watts) in the continuous mode are reached."* This is the "death-ray" concept of using a laser beam against targets such as an aircraft or missile warhead to produce such intense heating in a localized area that it burns a hole in the vehicle, weakening it structurally so that the craft is destroyed by aerodynamic forces.

The possibility of using the laser as a defense against ballistic missiles has prompted considerable interest because of its many potential advantages for overturning the pronounced advantage now enjoyed by the attacker. If an attacker knows approximately how many defensive ABM missiles are located at each site, he need only launch a few more ICBMs, or include decoys, to overwhelm and penetrate the defense. If, however, a laser ABM were feasible, the attacking warheads could be "zapped" as fast as they appeared since the intense beam of energy travels at the speed of light (186,000 miles per second).

But there are inherent difficulties in using a laser for ABM or air defense, even if sufficiently high power levels can be achieved. Some of the light passing through air is scattered in the process, and to reach ICBM intercept altitudes a laser beam must traverse many miles of atmosphere, thereby weakening it. Furthermore, an intensely powerful laser beam will heat up the air it penetrates, resulting in beam de-focusing which also dissipates some of the energy. Finally, an ICBM

---

* Laser power levels may be expressed in either of two ways, with a dramatic difference in the numbers involved. "Continuous, or average, power" is the output if the device delivers energy continuously. However, if the laser is operated intermittently, so that energy accumulates and is delivered in brief pulses, then the output power may be expressed as "peak power." Peak power is the average power divided by the duration of the pulse expressed in a fraction of a second. For example, a peak power of one billion watts would correspond to an average power of one thousand watts delivered in a pulse lasting only one-millionth of a second.

warhead is an extremely rugged vehicle, which has been designed to withstand the searing heat of atmospheric reentry. Whether it can already withstand the additional heat of a laser beam after attenuation by the atmosphere, or can be improved to withstand it, is a crucial point.

However, recent heavily censored Congressional testimony reveals evidence of progress in high-energy lasers and their possible application to ABM and air-defense systems. Perhaps the best evidence is the fast-rising funding being sought to support high-energy laser research and development. From a level of several million dollars a year in the mid-1960s, the Pentagon's high-energy laser effort has increased to approximately $30 million in fiscal 1970, and will run significantly higher in fiscal 1971.

During Congressional hearings on the fiscal 1971 defense budget in the spring of 1970, Dr. Eberhardt Rechtin, head of the Advanced Research Projects Agency, briefly discussed the Pentagon's high-power laser weapons program. He was then asked about his statement the previous year describing "a concept of employing an [deleted] as a defense against ballistic missiles." The censor deleted most of Dr. Rechtin's reply but left in his admonition that "it should be remembered that we are presently in the early conceptual design phase of the program with feasibility demonstration planned for [deleted] based on success at all upcoming milestones."

It is not certain at this point that the laser will ever become a useful weapon in ballistic-missile defense. But if it does, both the U.S. and USSR probably will want to deploy a laser ABM because it could be far more effective and much less costly than a system that requires defensive missiles. The system would probably use the same radars and merely substitute lasers for interceptor missiles.

Any decision in the U.S. to deploy a laser ABM must necessarily be debated in public, whereas the Soviets can decide and deploy in secrecy. The Russian lasers themselves could be installed underground, using construction no more conspicuous than that required for ordinary sewers or electrical power-distribution cables. If the Soviets were able to secretly deploy an extremely effective laser defense against American missiles and bombers, this could have a serious destabilizing effect on world affairs—even if only temporarily. It would put the USSR in a position where its "hawks" could propose a preemptive strike against the U.S.—confident that Russia could easily destroy any surviving American missiles and bombers sent to retaliate.

Admittedly, this sounds like the scenario for a *Dr. Strangelove* type of movie. It is advanced here only as a technical *possibility*, of which Soviet scientists are fully aware. If it sounds far-fetched, recall that only two decades ago the idea of obtaining pictures from reconnaissance satellites seemed an even more remote possibility.

In the early 1960s, one of the Pentagon's top scientists confided to me that he was extremely skeptical that the laser could ever be scaled up to power levels needed for "death ray"-type applications. But he added that the Pentagon ought to explore this possibility at a modest funding level in case this appraisal was wrong. Less than a decade later, another high-ranking Pentagon scientist speaks guardedly of plans for a feasibility demonstration of a weapon that had seemed all but impossible.

The lesson should be clear. In a precarious thermonuclear age, the U.S. needs to energetically explore the frontiers of science and technology so that the American intelligence community can know which things are technically possible for a potential adversary. We must also have an active program to continuously upgrade the capabilities of American reconnaissance satellites so that photo-analysts can readily detect, and understand, any signs that a potential adversary has developed a new type of weapon that could suddenly break the existing thermonuclear stalemate.

22

FATEFUL DECISION, "GILPATRIC'S PRINCIPLE" AND THE FUTURE

A heated debate must certainly have preceded the fateful decision of Soviet leaders to accept the principle of freedom of observation from space and not to attempt to destroy American reconnaissance satellites. Less than a decade earlier, Soviet leaders had rejected Eisenhower's Open Skies proposals, convinced that Russia had far more to lose than gain by forfeiting its cherished secrecy.

In late 1958, when Soviet and American technical experts met in Geneva to discuss possible ways of preventing a surprise attack, there was no sign of any change in the Russian attitude. If anything, it seemed more firmly entrenched. The chief American negotiator at Geneva, William C. Foster, later observed during Congressional hearings: "We were much impressed by the importance which the Soviet representatives attach to secrecy as a military asset. In effect, they seem to believe it enables them to possess a form of 'hardening' of their bases which we do not have. Thus they regard any encroachment upon this secrecy as a unilateral disarmament step on their part which must be compensated for by other measures." Then in May 1960 the shooting down of the Powers U-2 and the Soviet reaction to the incident demonstrated that the Russians would not tolerate any penetration of the Iron Curtain if they could prevent it.

In the late 1950s, when Russia first learned of America's plans to develop photo-reconnaissance satellites, and heard American scientists discuss the general capabilities of space-borne observation at Geneva in 1958, there must have been those in high places who argued that Russia should quickly develop weapons to destroy these new probing eyes in space— the successors to the despised U-2.

Similar views were expressed in the U.S. during the early 1960s by those who urged that weapons be developed to "neutralize unfriendly satellites" if that should become necessary. By the mid-1960s, the U.S. had modified a few Thor IRBMs and Nike-Zeus ABMs to function as satellite destroyers if the Russians should ever violate the treaty against putting nuclear weapons in orbit. Certainly the Russians could have done the same, perhaps even earlier.

Possibly the initial Soviet decision to accept spaceborne reconnaissance resulted from the recognition that by the time Russia could develop an effective anti-satellite weapon, the most vital Soviet secret would already be known to the U.S., i.e. how few ICBMs the Russians had actually built. Probably Russian leaders themselves were anxious to obtain their own satellite photos to be certain that the U.S. really had built a massive missile arsenal before committing the USSR to an extremely costly effort to catch up.

Perhaps the initial Soviet decision also was influenced by the belated recognition that Russian secrecy, in combination with the relative openness of the U.S. defense activities, had backfired. The Iron Curtain had been so effective that U.S. defense programs had to be based on intelligence estimates which were shaped by those manifestations of missile power that the Russians chose to reveal in their space spectaculars and missile tests. The resulting U.S. fears of massive Soviet missile attack had led to the development of the submarine-launched Polaris and the hardened, quick-reaction Minuteman. The U.S. believed these costly designs were needed to protect against a surprise Russian attack and assure survival of a sufficient retaliatory force to prevent such an attack.

But the net result, as seen through Russian eyes, was to place the U.S. in a position where, if it chose, it could launch a devastating preemptive strike against the USSR.

The ironic consequences for the USSR of Soviet secrecy and American openness were succinctly expressed by Deputy Defense Secretary Roswell Gilpatric, on August 13, 1962, during a seminar at the Air Force Academy in Colorado Springs: "The Soviets are forced to work very hard to keep up with what *they know* we are doing in order to keep up with what *we think* they are doing," Gilpatric said.

Whatever the combination of factors that prompted the Kremlin decision in the early 1960s, it is interesting to consider what might have happened if the Russians had decided to try to destroy American reconnaissance satellites. Space would then have become a battleground in which no satellite, American or Russian, would have been safe from destruction. It would have been difficult to tell which spacecraft were innocent weather-observation or scientific satellites and which were performing strategic reconnaissance. Even manned spacecraft would be exposed to the risk of destruction since the Russians might naturally presume that the astronauts were using cameras for reconnaissance purposes.

Whether a Soviet decision to try to block American spaceborne reconnaissance would have been completely effective, or would merely have greatly reduced the amount of intelligence that could be obtained, one thing seems certain. Once again the U.S. would have been forced to rely heavily on intelligence estimates of Soviet missile strength, and once again "Gilpatric's Principle" would have prevailed.

The U.S. would not have cut back and limited the number of Minuteman missiles to 1,000, as was done in the mid-1960s. Instead, it almost certainly would have gone on to build 2,500, or more, as recommended by the USAF in the early 1960s. Instead of limiting the Polaris submarine fleet to forty-one vessels, today the Navy probably would have close to a hundred. There would be no intellectual debate over whether or not to deploy an ABM. The only question would be how large a system should be constructed and there would be strong public pressure to extend its coverage. Uncertainty over Russian military strength would probably prompt the U.S. to spend $10–$15 billion a year more than it now devotes to defense expenditures—an annual increase comparable to the total cost of the reconnaissance-satellite program since its inception. The Russians, seeing the massive U.S. effort, would be racing to keep pace, thereby making U.S. intelligence estimates self-fulfilling.

A mysterious space experiment, conducted by the Soviets in October 1968, using Cosmos-248, Cosmos-249 and Cosmos-252, suggests that the Russians recently may have developed a satellite designed to destroy other spacecraft. Cosmos-249 and Cosmos-252 mysteriously fragmented into dozens of pieces, apparently when they came near Cosmos-248. The tests were conducted at an orbital inclination of 63 degrees, never before used by the Russians, as if they expected fragmentation and wanted to keep the debris out of orbital inclinations used for other Soviet spacecraft. Two years later, in October of 1970, the Russians conducted what appears to be an identical experiment, using Cosmos-373, -374 and -375, in which the last two have disintegrated into dozens of fragments, apparently while in the vicinity of Cosmos-373. Full details are not available because the Pentagon has clamped a tight security lid on the subject.

If these two incidents do indeed involve tests of a "killer-satellite," it does not necessarily mean that the Russians intend to reverse their previous policy and will soon try to destroy U.S. reconnaissance satellites. The Soviets may have developed this type of anti-satellite capability to use if the U.S. should try to place nuclear bombs in orbit in violation of the treaty banning such weapons in space. (The U.S. has no plans for such orbital weapons.)

But the implications could be more ominous. The Pentagon's security measures on all data involving the most recent Russian tests indicate that there is concern at high U.S. government levels.

During SALT discussions with the Soviets, the U.S. will press for an agreement that neither nation will interfere with the reconnaissance satellites of the other, since they would be needed to police any arms-control treaty, according to Chalmers M. Roberts, writing in *The Washington Post* on November 1, 1970. (The article appeared just after the

Cosmos-373, -374, -375 tests, but the U.S. decision to press for such an understanding pre-dated the most recent Soviet experiments.)

Today, from the vantage point of hindsight, there are some who criticize the U.S. for "overreacting" to the Missile Gap. If the threat seemed larger in the late 1950s than it truly was, this was by Soviet intent, as evidenced by the frequent boasts and rocket-rattling of Russian leaders. Eisenhower's final State of the Union message, cautioning against excessive influence by the "military-industrial complex," is frequently cited today by those who now suggest that the Missile Gap was a fiction perpetuated for self-serving ends. Forgotten is the Eisenhower warning of the grave new threat of surprise ICBM attack which he sounded only four days later.

The Missile Gap was fact, not fantasy. It would have persisted much longer, well into the 1960s, if the U.S. had not launched its crash program when it did, in early 1954. If the U.S. had waited until mid-1957, when the Russians saw fit to unveil their new "ultimate weapon" and began to threaten its use, President Kennedy might not have been able to withstand Khrushchev's Berlin ultimatum. The President would have had to make crucial decisions without the benefit of reconnaissance-satellite photos. Certainly the Soviets would have used missile blackmail for geopolitical advantage elsewhere, and the two countries might have bumbled into World War III as a result.

It would have required a person with extraordinary vision in the early 1950s to foresee that the marriage of two terrifying new weapons of mass destruction—the H-bomb and the ICBM—would produce a satellite offspring that could so drastically change, and stabilize, strategic relations between the U.S. and the Soviet Union. It would have required amazing prescience in mid-1955, after the Russians had so vigorously rejected Eisenhower's Open Skies proposal, to have predicted that the Soviets would reverse their position, at least tacitly, within the decade.

The record shows that in early 1955 Colonel Richard S. Leghorn, an Air Force reconnaissance specialist who later became one of the founders of Itek Corp., proposed that the

United Nations consider developing surveillance satellites.
And on February 4, 1958, Senator Hubert H. Humphrey, during Senate debate, proposed that the UN undertake a reconnaissance-satellite effort: "A satellite of this nature would
impress all nations that no longer are national borders and
countries sacrosanct," he said. "It would be a vivid example
of internationalism which by its very existence would require
the creation of new concepts of international law and order."*

But the record also shows that Prime Minister Churchill's
scientific adviser, Dr. F. A. Lindemann, dismissed the import
of the German V-2 in late 1944 with these words: "Although
rockets may play a considerable tactical role as long-range
barrage artillery behind the lines, at 20, 30 or even 50 miles,
I am very doubtful of their strategic value." A similar type of
appraisal came in the late 1940s from America's highly respected Dr. Vannevar Bush. The record also shows the grave
concern of the brilliant Dr. J. Robert Oppenheimer, and many
of his associates, that no good could possibly come from the
development of the H-bomb. And the record shows that
shortly after the first Sputnik went into orbit, marking the
dawn of the Space Age, one of President Eisenhower's top
aides called the Russian achievement "a silly bauble."

There are those today who argue that the U.S. and the
USSR are much less secure than they were a decade ago,
despite the billions of dollars spent for ballistic missiles and
thermonuclear warheads. This ignores the fact that without
the H-bomb, the ICBM and IRBM would not have been
viable weapons. And without these powerful rockets, there
would be no reconnaissance or surveillance satellites. Without
spaceborne reconnaissance, the U.S. and the USSR would be
engaged in a far more massive and costly arms race, and
there would be no SALT discussions under way.

Livy, in his *Annals of the Roman People,* expressed it well,
nearly two thousand years ago: "The greatest fear is of the
unknown." In an age when thermonuclear devastation can

* More recently, in remarks delivered on the Senate floor on March
25, 1971, Humphrey urged a freeze on ABM deployment by the U.S.
and the USSR, adding that compliance "could be policed through the
satellite reconnaissance systems of each side . . ."

cover intercontinental distances in thirty minutes, the secret sentries in space have done much to ease this "greatest fear."

But the ingenious reconnaissance and surveillance satellites can only provide *information.* If government leaders fail to use this information wisely, if they release driblets of satellite-derived information only when it serves their political purpose of the moment, the American public may fail to react at a crucial time when strong reaction is called for, and these powerful weapons for peace will have been wasted.

Soviet leaders, including Russian "hawks," must fully recognize that the Iron Curtain no longer exists and show restraint in expanding their inventory of ICBMs. If they don't, the prying eyes in orbit will serve to spark an ever-expanding arms race, instead of controlling it. And if the Soviets are tempted to use their "killer-satellite" capability to destroy American reconnaissance or early-warning satellites, they should consider the consequences. It will turn space into a battleground, precipitate a still more costly arms race and return the world to the perilous days of the late 1950s.

INDEX

INDEX

A

ABM (Anti-Ballistic Missile), xii, 70, 117, 188, 194, 196–97, 199, 200, 209, 212–13, 216
AEC (Atomic Energy Commission), 13–15, 186
Aerojet-General Corp., 182
Agena (second-stage and satellite), 83, 88, 90, 92–93, 102, 105, 123, 131, 137, 153, 174–75. *See also* Atlas/Agena, Thor/Agena
Airborne Alert (SAC bombers), 38, 58–59
Airborne Instruments Laboratory div., Cutler-Hammer Corp., 192
Air Force, *see* U.S. Air Force
Air Force/Space Digest magazine, 78n
Alice Springs, Australia, ground station, 182
All-American Engineering Co., 91, 93
Allen, George V., 48
Alsop, Joseph, 65–67, 100–1, 107
Alsop, Stewart, 114
Andrews Air Force Base, 168
Arnold Engineering Center, 96
Associated Press, 199, 200
Atlas/Agena launch vehicle, 90, 92, 102, 104, 106, 111, 130, 138–39, 162, 174–75, 179–81, 192–93
Atlas ICBM, 14, 16–17, 19, 23, 28, 31–32, 34–35, 39–42, 45, 49, 52–53, 55, 59, 68–69, 83, 102, 107, 115

ABOUT THE AUTHOR

PHILIP J. KLASS, senior avionics editor for *Aviation Week & Space Technology* magazine, graduated from Iowa State University in 1941 with a B.S. in electrical engineering and "an intense dislike for writing of any kind." For ten years he practiced engineering with General Electric in that company's avionics (aviation electronics) activities while slowly but surely gravitating toward technical journalism. In 1952 he joined *Aviation Week*. Mr. Klass's previous book, *UFOs— Identified* (Random House, 1968), debunked the idea that the earth is playing unwitting host to spaceships from other worlds.

Mr. Klass, who lives in Washington, is an enthusiastic skier and a casual Potomac River sailor. Formerly an ardent Civil War buff, he has designed and built animated electronic displays that tell the story, complete with sound effects, of major battles at Gettysburg and Antietam. He is a senior member of the Institute of Electrical and Electronics Engineers, a member of the American Institute of Aeronautics and Astronautics, the American Association for the Advancement of Science, the Aviation/Space Writers Association and the National Press Club.